Integrated Materials Handling in Manufacturing
Willi Müller

Springer-Verlag
Berlin Heidelberg GmbH 1985

British Library Cataloguing Publication Data

Müller, Willi
   Integrated materials handling in manufacturing
   1. Materials handling
   I. Title
   658.7′81   TS 180

ISBN 978-3-662-12094-1   ISBN 978-3-662-12092-7 (eBook)
DOI 10.1007/978-3-662-12092-7

Translated from the German by Sue Greener
Phototypeset by Wordsmith Graphics Ltd, Street, Somerset

Acknowledgements

MANY people have helped with the preparation of this book. In particular thanks are due to W. Eversheim, Professor of Production Systems at the Rhein Westphalia College of Technology in Aachen for his kind support, generous help and valuable suggestions; Professor M. Weck, Professor H. Rake and Professor F. Hildebrandt, for their general comments; and Dr K.-W. Witte and Mr G. Burkhart for their suggestions and comments during preparation of the manuscript.

In addition, thanks are due to colleagues at the College of Technology who have given support through their advice and readiness to help – in particular Mrs D. Tebbe, Mr B. Stankowski, Mr M. Cassel, Mr H. Gehlen, Mr M. Körner and Mr J. Sauer.

Finally, special thanks are due to my wife, who through her understanding and her willingness provided the basic support necessary for this book to be produced.

Contents

Chapter One
Introduction

MANUFACTURING in mechanical engineering enterprises is characterised by fundamental changes both inside and outside the organisation[1].

Inside the organisation, account must be taken of increasing cost pressures in wages, materials, energy and stocks, all of which necessitate making the most effective use possible of manufacturing capacity. Outside the organisation, changes in consumer demand take place. These stem from increasing domestic and foreign competition which result in an increase in the volume of deliveries, a proliferation of products and a related shortening of product life-cycles. These give rise to wider component ranges involving smaller quantities which, because of increasingly necessary changes in materials and despite higher stocks, still lead to longer throughput times.

These trends impose a continuous pressure on businesses to take steps to rationalise both organisational and technical procedures.

When considering the introduction of new technology and raw materials or the re-organisation of labour-intensive manufacturing plants, considerable weight has been given up to now to the automation of production processes. As shown in Fig. 1, which is an analysis of working hours of manufacturing units, successful rationalisation has been achieved already in this way; this has brought a relatively high and cost-effective utilisation of the manufacturing facilities related to the planned operating time[2]. The reasons for the current low performance are to be found on the one hand in the occurrence of technical, organisational and personnel-related interruptions and on the other hand in mainly personnel-related stoppages in unused shifts and during holidays and weekends[3].

Opportunities for rationalisation give way in the first instance to an acceleration of production processes. These measures at present, however, come up against a range of technological and mechanical

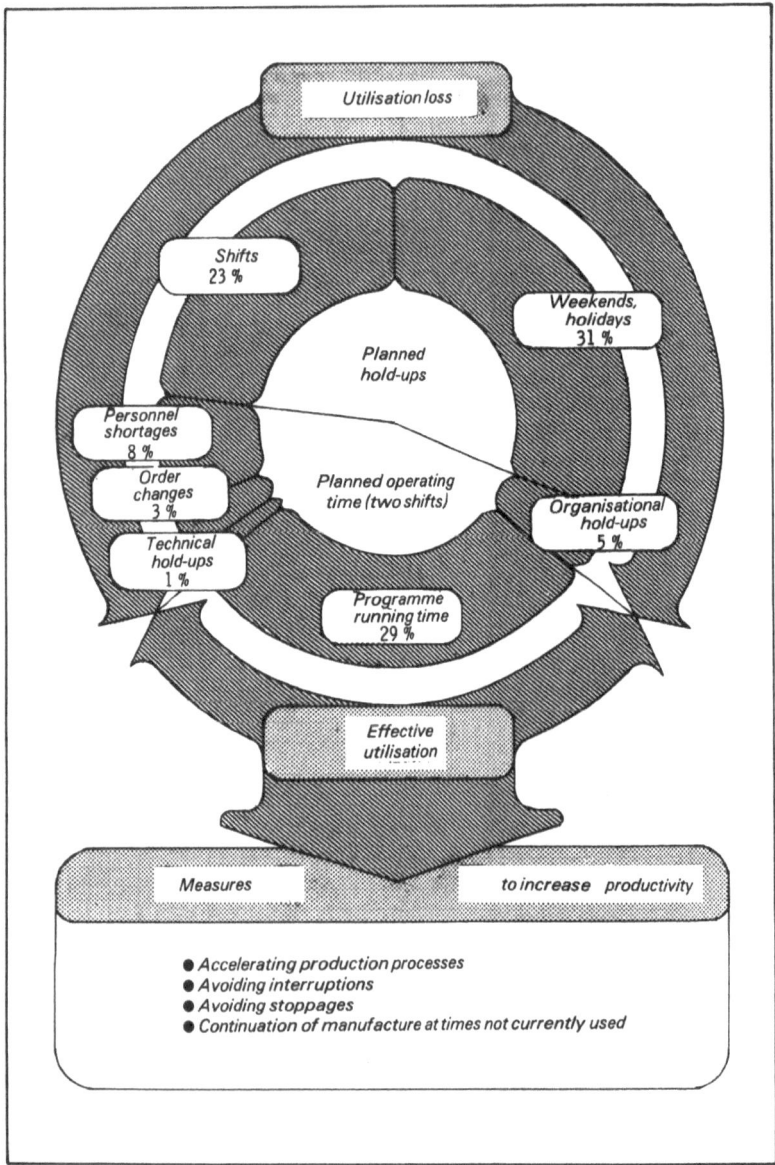

Fig. 1: General measures relating to productivity increases in production plants

difficulties (for example, the condition of the tools, the static and dynamic behaviour of the machine tools) but these will be overcome in individual cases. However, they currently allow no significant increases in productivity.

Beyond that, the possibility exists for avoiding stoppages within the production processes which are still necessary in existing operations, such as those that take place through component and tooling change procedures or task alteration. It is assumed here that on the one hand the plant permits at least partial completion of tooling processes and tools or component replacement as running changes. On the other hand, procedures for start-up and personnel operation must be used which guarantee the smooth-running operation of the production plant[4].

To achieve a further increase in plant utilisation, suitable inspection systems can be instituted and these can help the production process to be more carefully controlled and thus avoid stoppages[5].

Further possibilities for rationalisation are available by carrying out production at times when the plant is not currently being used. However, that involves comprehensive automation of all production processes, as during these times, the necessary personnel are either absent or are only available in limited numbers.

Automation under these circumstances must be accompanied by organisational changes. However, they, in turn, assume that technical advances in the production process do exist.

Chapter Two
Analysing problems and setting objectives

A N INVESTIGATION of the processes occurring within produc-tion shows that they are constructed from a repeated cycle of storage, conveying and manufacturing processes. Handling processes are involved in the transfer of components from one production facility to another. These processes serve for picking up components from one of these facilities and passing them on in an appropriate condition to another for the next stage of production (Fig. 2). In this context, production facilities are defined as not only being manufacturing but storage and transport functions as well.

Efforts to date in the automation of production processes have often necessitated independent equipment and machines[6-11]. While the automation of conveying, storage and manufacturing facilities can be considered as the latest technology (e.g. NC turning machines, high racking stores, driverless trucks, and so on), the operation of automatic handling facilities achieves only qualified success[12].

One of the reasons for this is the defective mutual adaptation of individual storage, conveying and manufacturing facilities, through which a higher technical and financial outlay is caused in both the supply and operation of this kind of facility (Fig. 3).

The result is that, in addition to the complex technology required for the handling equipment and its peripherals, the costs of automating handling processes in production often exclude an economic solution.

It is therefore the aim of this study to integrate the handling processes into facilities for storage, conveying and manufacturing so that no independent, universally applicable handling equipment is required for the transfer of components between individual production units.

To this end, engineering concepts will be developed and presented. These will help in the execution of handling processes among storage, conveying and manufacturing facilities.

In most medium-sized engineering firms this type of production

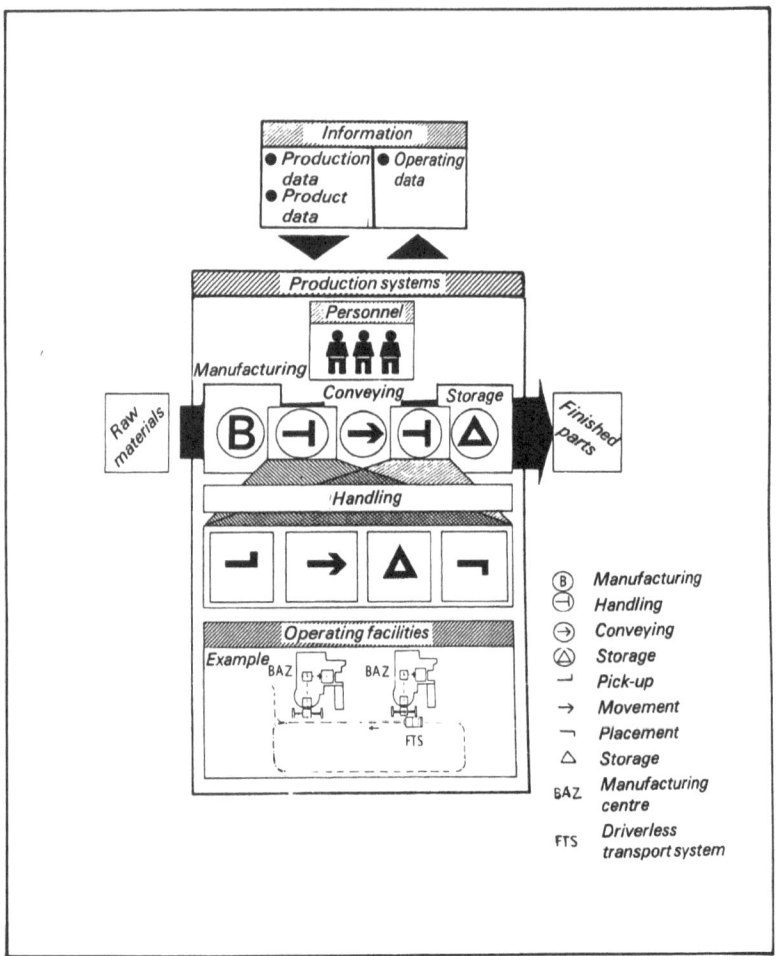

Fig. 2: Materials handling in the production process

facility using integrated handling operations must have the following set of conditions:

- Through the integration of handling facilities in production systems, resources for technically straightforward automation should be made available.
- The system should allow the operation of manufacturing facilities

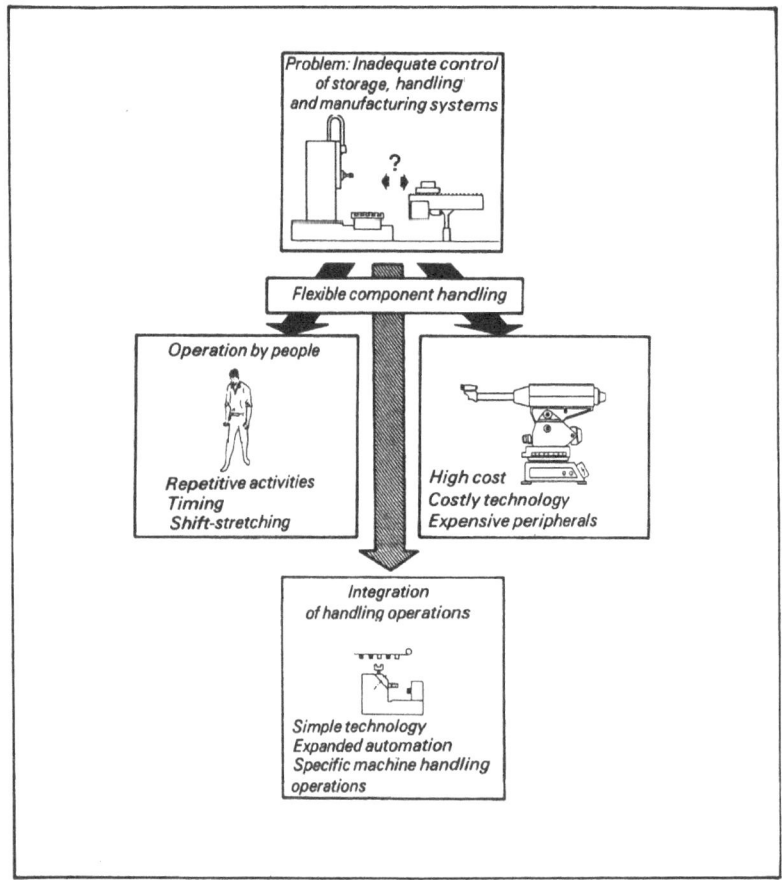

Fig. 3: The necessity for integrating handling operations in production

in previously unused times, as well as a reduction in operating stoppages.
• The integration of handling facilities in production plants should guarantee the company a far-reaching flexibility in changing production conditions.

Furthermore, production plants with integrated handling operations should help create improvements in the working conditions for employees.

The focus of this study lies in the range of components of average

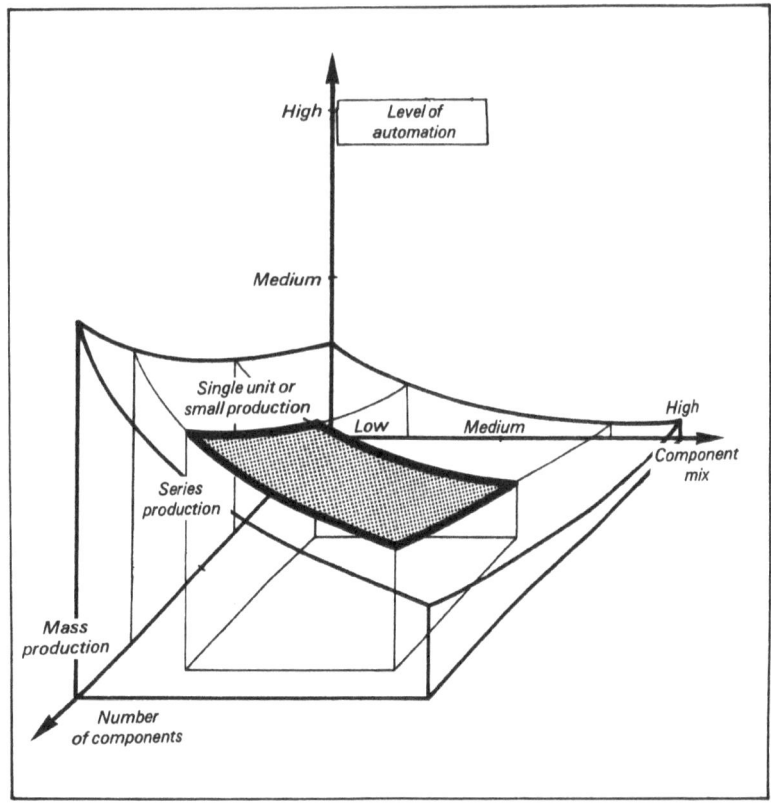

Fig. 4: Operational scope of production facilities with integrated handling operations

numbers (approx. 20-500 components) and average component mix, as might be found, for example, in firms with multiple product, medium-sized production runs (Fig. 4).

The investigation is concerned with production systems involving metal-cutting and shaping operations which, according to their level of automation, extend from highly automated production systems, such as transfer lines and flexible manufacturing systems, to labour-intensive production in which a majority of manufacturing processes are not easily automated.

Production systems will be dealt with which have not yet been thoroughly researched (as is the case with transfer lines and flexible production systems) but which can be automated; this is especially so in terms of material handling using currently available techniques. In particular, turning machines will be considered as well as drilling and milling machine tools (i.e. cutting tools involving rotary motion). The results produced and the procedures adopted can, however, generally be used for other production systems (e.g. planing machines, finishing machines, presses, and so on).

The study does not concern itself with the handling of tools, peripheral equipment and measuring instruments in the production process, but deals with fundamentally similar processes for materials handling.

However, because of differing geometry, flow paths, time factors, and so on, handling of the above mentioned facilities represents at any given time an individual set of problems.

Chapter Three
Integration in the context of technical investment planning

THE SUBJECT of the integration of handling operations within production plants involves in the first instance the provision of concepts for storage, conveying and manufacturing facilities without direct reference to a defined production task. Thus, by supplementary measures, the production facilities can be placed in a position to perform the necessary processes for materials handling.

The integration of handling operations in production facilities thus creates the basis for the planning of production systems, in which this kind of production facility with an integrated handling operation will be used. The integration of handling operations in production systems is an integral part of technical investment planning and thus belongs to the detailed planning stage of storage, conveying and manufacturing facilities (Fig. 5)[13-18].

The starting point in any investment analysis for planning production facilities having an integrated handling operations is the production task. This is determined by manufacturing requirements and materials flow. From these two dimensions can be calculated the size and output capacity of the storage, conveying and manufacturing facilities. In this way the characteristics of these production units can be ascertained; in particular, the type, number, and layout of machine tools, as well as the means of transferring components and the inspection or measuring techniques.

The integration of handling operations builds on to these results, in that the production facilities will be adjusted in terms of one another and linked together from a technical handling standpoint.

The purpose of the following chapter is to investigate both aspects of the integration of handling operations in production plants, bearing in mind the provision of new engineering concepts together with the planning of the operation of these concepts in production systems.

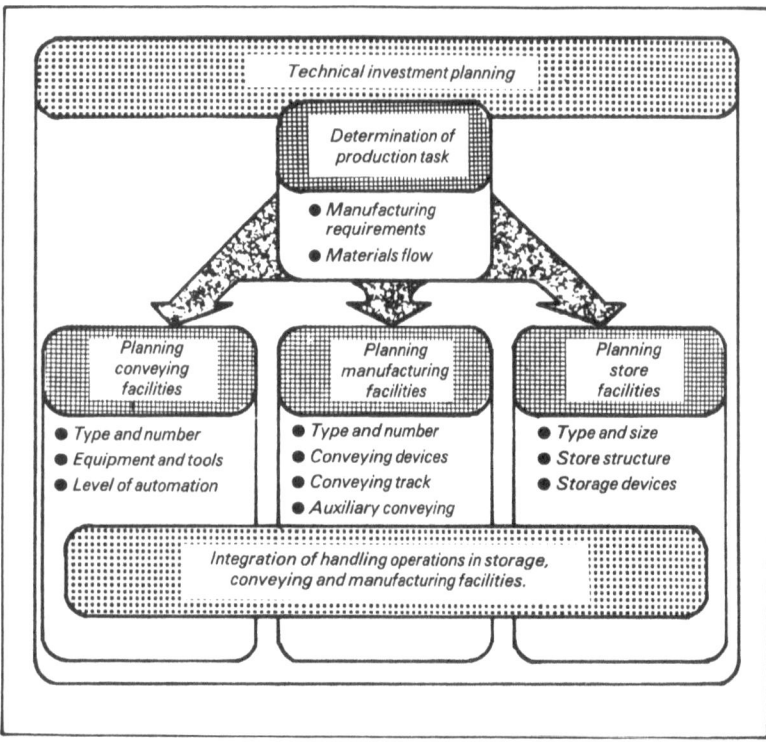

Fig. 5: The integration of handling operations in storage, conveying and manufacturing facilities as an integral part of technical investment planning

Chapter Four
Using the latest technology

WORK to date in the field of handling technology roughly depends on the volume of throughput planned for the operation. In mass production, handling equipment has been installed to link together manufacturing facilities, and these are adjusted to the specific needs of the manufacturing facilities and the components or materials to be handled[19]. Thus, either special equipment would be developed or the handling facilities would be generally integrated into the manufacturing facilities, especially when very short cycle time manufacturing processes must be linked together[20-22].

For larger throughputs and, depending on the size of the components, handling facilities would be installed which could carry out complex handling processes. So for very small components, handling facilities would be found which could undertake a large number of ordering, sorting, transferring and feeding tasks. Examples are vibratory bowl feeders, centrifugal feeders, elevators, and so on[23,24].

With larger products, operations of this kind are found only in the production field. Here, rough preparatory processing is undertaken (e.g. forging, casting), since in these handling facilities, the products are usually handled as bulk materials. In that way, in production areas with higher demands for quality, inadmissible deformation and damage to the product would be caused[25].

In this kind of production area, handling machines such as insertion equipment, transferring facilities, swivelling and turning equipment come into operation to handle the products as single items. In this way, up to now only simple, repetitive handling processes can be carried out.

Complex handling processes, such as movements in several directions, can be achieved only by several inter-connected handling facilities. While small components in mass production often appear in random order (in which condition they are handled automatically) larger components usually must be arranged manually before they can be taken over by automatic handling facilities.

Developments in the field of single components production and medium production runs have usually involved the provision of independent equipment [26-30]. Corresponding to the higher demands for flexibility made of this equipment, the question here is one of complex facilities which can carry out the most varied handling operations. Because of the high costs involved in the development of this type of equipment, manufacturers have made sure that currently available flexible handling equipment usually has various universal features which technically can be used in the most varied applications. This is to ensure in addition a greater potential market for this new technology.

Specially designed handling equipment to meet the requirements of applications with only single or medium run production are only rarely found.

For flexible handling equipment compatible with manufacturing facilities, it is a matter either of self-standing or portal equipment or of hardware which can be attached to the manufacturing equipment. The latter are often attached to turning machines. Here, because of the simple handling operations, placing the symmetrically rotating component into a stationary position means that costly mechanisms to carry out handling processes largely can be avoided. With free-standing handling equipment it is often necessary to have complex multi-axis equipment because of incompatibility between handling apparatus and machine tools[31-34].

In just a few cases the configuration of machine tools for single and medium production runs is compatible with the requirements of automatic handling apparatus. Above all the spindles of turning machines are set up offering supplementary movements[35,36].

A particular example within automatic materials handling is depicted by materials handling with the help of pallets and pallet-change systems. Here uniform handling conditions can be created through the use of pallets, thus ensuring that technically simple handling equipment may be used. It is essential in this case that the components are fixed, usually manually, onto the pallets[37,38].

To be precise, a distinction is made between the handling of components and the handling of pallets – provision, storage, transport, delivery, and so on. In the scope of this study, care will be taken to consider current pallet change methods, pallet pools, etc., not as free-standing handling equipment but as integral parts of the manufacturing facility.

With regard to the automation of handling processes in storage facilities, back-up equipment has been developed which can handle the goods to be stored using standardised component carriers (storage pallets)[39]. Handling is confined therefore to relatively simple processes such as lifting/lowering, shifting and turning and is related to the total number of components which can be placed on this form of transport. Also, storage facilities in the form of rotary store installations were developed. Here the necessary movements of goods within the store will take the place of handling processes[40]. This type of rotary store is used in connection with flexible handling capabilities for manufacturing facilities [41].

Handling processes for conveying methods are achieved with simple turning, lifting, lowering or shifting equipment[42-45]. Similarly, in automated stores, standardised component carriers are used.

Chapter Five
Developing alternatives

THE SYSTEM shown in Fig. 6 is seen as a development of the concepts for storage, conveying and manufacturing facilities with integrated handling operations.

For the technical achievement of handling processes, the limiting conditions under which these processes can be carried out are important. Therefore, the influence of production facilities on automatic handling was investigated first. Above all, the requirements of the production facilities, both in terms of movement and time constraints on handling operations, were determined.

Parallel to this, the individual processes in storage, conveying, manufacturing and handling of materials were investigated. In this way, a functional model was produced to help determine the processes in the production operation.

The result of this stage of the work is an overview of functions which it is necessary to automate through the integration of handling processes in production facilities.

On the basis of requirements in product handling and after establishing the functional model, alternative solutions were developed for production facilities with integrated handling operations. In addition, the principal sequences of operations could be identified. Technical solutions for the realisation of handling processes in storage, conveying and manufacturing facilities could be assigned to these sequences.

Studying the effects of storage, conveying and manufacturing facilities on automatic handling

The requirements which are placed on an automatic handling system by storage, conveying and manufacturing facilities follow from the individual characteristics of their components.

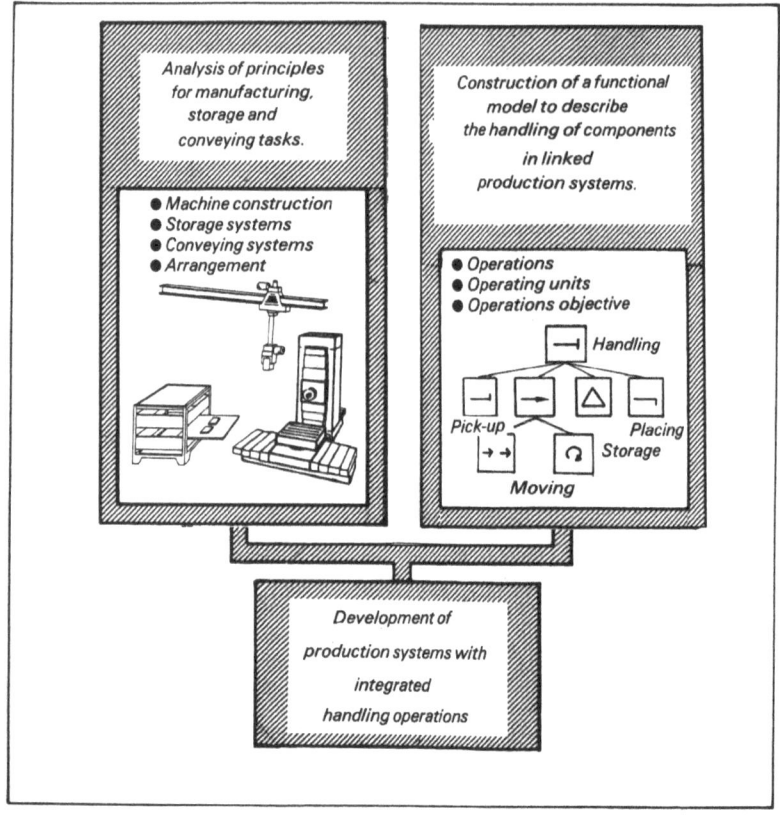

Fig. 6: System for the determination of alternatives for the integration of handling operations in production plants

The relevant components of a store are listed in Fig. 7[46]. Under the heading of 'support apparatus' are included store-related conveying systems, which take one or several stored components from their place and move them to a waiting or idle station or to another conveying system. This type of support equipment is well-known for stores with high racking[47]. Support in stores is provided by component carriers in the form of transport pallets, bins or specially constructed items which facilitate component storage.

The relevant components of the conveying system can be similarly defined. The conveying techniques, for example, consist of all the equipment required for carrying and taking components along a given path. Individual conveyors, such as forklifts, driverless floor conveyors,

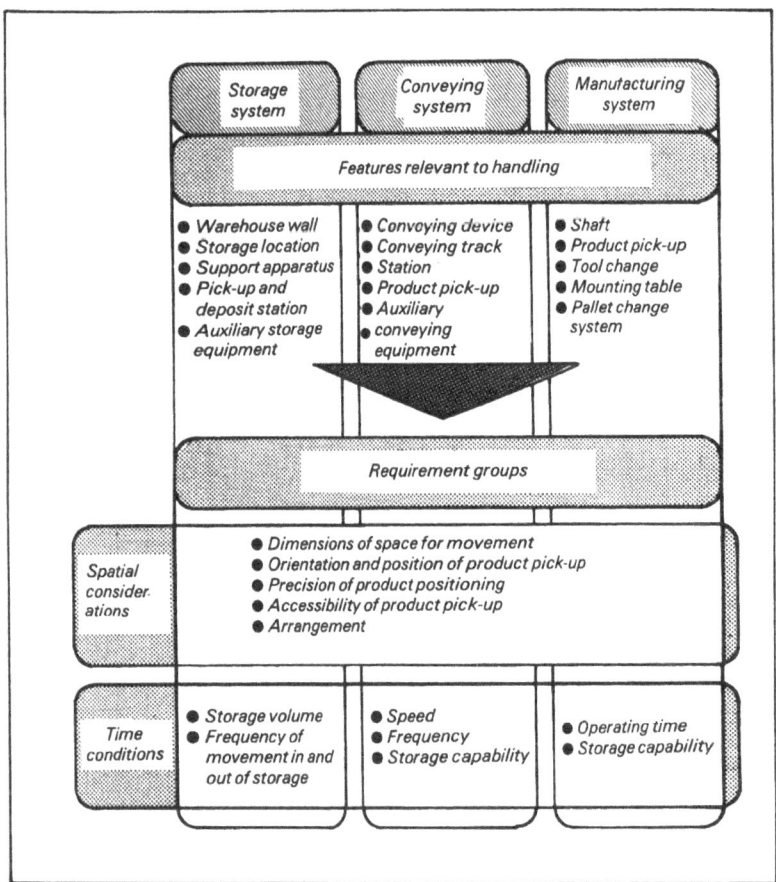

Fig. 7: Influence of storage, conveying and manufacturing facilities of automatic handling

and monorail overhead conveyors, correspond exactly to the vehicles; whilst continuous conveyors, such as drag-chain conveyors or slat conveyors, come under the heading 'means of conveying'. These are understood as the single hanger which takes single items along the level or up inclines[48-50].

The stations in conveying systems correspond to the waiting stations in storage systems, since, at these, components can be transferred to other facilities in the production system.

Loading surfaces, brackets, hooks and so on are designated as 'component pick-up points', whilst component carriers are identified as an auxiliary means of conveying. Sometimes, in the case of storage and conveying processes which follow one another, the same storage or conveying auxiliary support will be used[51].

In manufacturing systems, all facilities for the transmission of working energy are understood by the term 'shaft'. In contrast to machines for milling or drilling which have the shaft picking up the tool, especially in turning machines the clamping device for components (e.g. triple jaw chuck) is part of the shaft. In the framework of this study, the term 'component pick-up' is understood to include the steady rests for turning machines and the largely component-specific attachments for milling machines. Mounting tables are machine components which serve in milling machines for the placement of attachments.

Pallet changers are devices which can carry out handling processes with the aid of standardised product carriers (pallets) to supply manufacturing facilities.

The requirements of automating handling processes derive from the features of production facilities which are relevant to handling; these requirements result from the present space and time conditions of the production facilities[52-54]. When investigating spatial demands the space available for movement in the handling of products will be considered first. Space for movement within storage facilities will be limited by the dimensions of the store compartments. This does not apply to level stores just as for most conveying systems where the open construction means that space for movement is limited only by the distance to other components or other production units.

The space for movement at manufacturing facilities will be dictated mainly by the workspace dimensions. In Fig. 8 the distribution of workspace dimensions of turning machines is represented. As the comparison of the diameter of turning machines with the component diameter of an actual component range shows, the space conditions of turning machines in general are set up so that sufficient room remains for the equipment to handle the components[55]. It is thus assumed that when the machine is selected, appropriate dimensional margins are planned, along with correspondingly higher financial cost.

For the further estimation of space conditions in production facilities, next to space for movement, it is important to determine the orientation and position of component pick-up. In storage facilities, the orientation and position of component pick-up (store compartments) are determined by the dimensions and construction of the store. It may be rectangular, cubic, circular or cylindrical in shape. Component pick-up points will lie accordingly in a rectangular or circular arrangement[56,57].

With conveying facilities, the orientation and position of component pick-up points vary in relation to the conveying path and they change

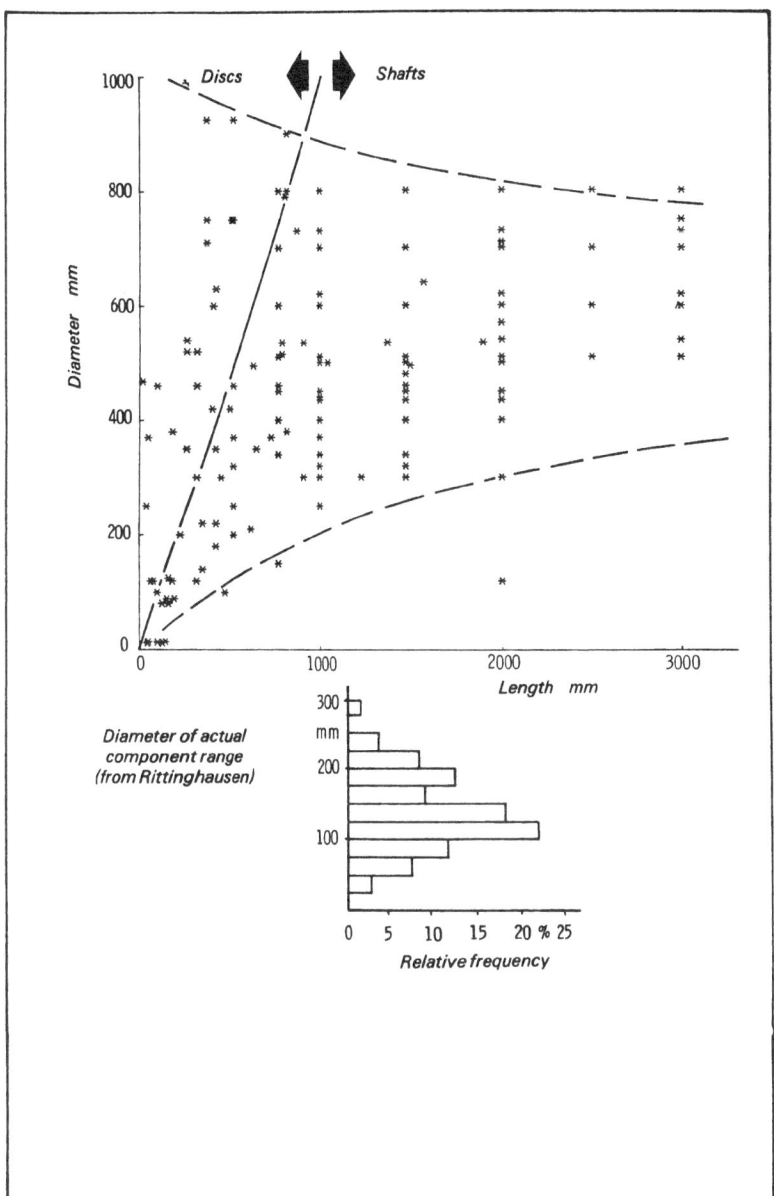

Fig. 8: Distribution of workspace dimensions in turning machines

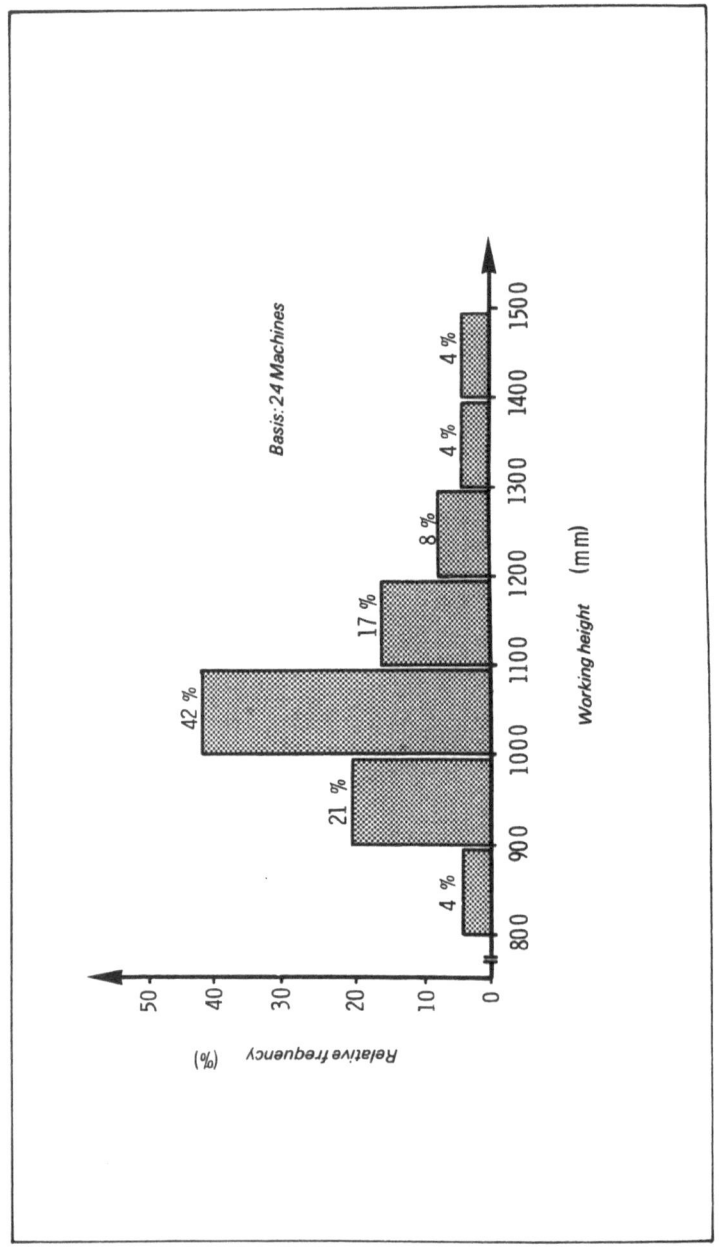

Fig. 9: Frequency distribution of working heights of turning machines

according to the direction of curvature. In addition, for conveying, there exists lifting, lowering, swivelling and shifting apparatus with which the orientation and position of component pick-up can be varied.

With manufacturing facilities, the orientation and position of component pick-up is determined also by dimensions and construction, together with the distribution of movement axes. Fig. 9 shows the frequency distribution of work heights, i.e. for turning machines with a horizontal shaft the position of the shaft over the floor[58]. Here the height of the shaft depends on maximum diameter and on critical constructional sizes, e.g. the power configuration and that of the machine bed.

The requirements of handling process automation, which are dictated by space, depend not only on the previously mentioned critical sizes but on the precision of the component position and orientation. With turning machines and milling machines the position and orientation of the component is generally defined by the output position, whereas with component insertion, the two dimensions can be subject to tolerances. The allowable tolerances are determined by the configuration of component pick-up points (Figs. 10 and 11)[59].

The various component pick-ups are differentiated first of all by the need for precise positioning and orientation in individual axes. Furthermore, the number of attachment elements which will make a straight-line feed more difficult has to be considered[60].

The tolerance range for component placement observed by a handling system will be mainly determined by the position and number of carrying blocks as well as by the clamping of attachments. Here the tolerance of carrying blocks is equal to 0, as these must be achieved exactly. On the other hand, the tolerance range for carrying a plain surface, in all directions parallel with that surface, is great.

In an analysis of the accessibility of production facilities, the possibilities of straight-line feed of components from a fixed direction is investigated (Fig. 12). Deviating from existing systems of axis specification, no distinguishing characteristic of axes is selected with reference to the axes of manufacturing facilities. The machine-neutral designation of feeding directions should make possible a comparison of different manufacturing facilities with regard to their accessibility. In storage facilities – except level stores – access is usually only from one main direction. However, conveying facilities are usually accessible from various directions.

In manufacturing facilities, accessibility is restricted by construction as well as equipment such as tools and attachments. As an analysis of accessibility to single shaft-turning machines shows, turning machines are easily accessible from a horizontal direction.

However, an exception can be found with horizontal bed machines.

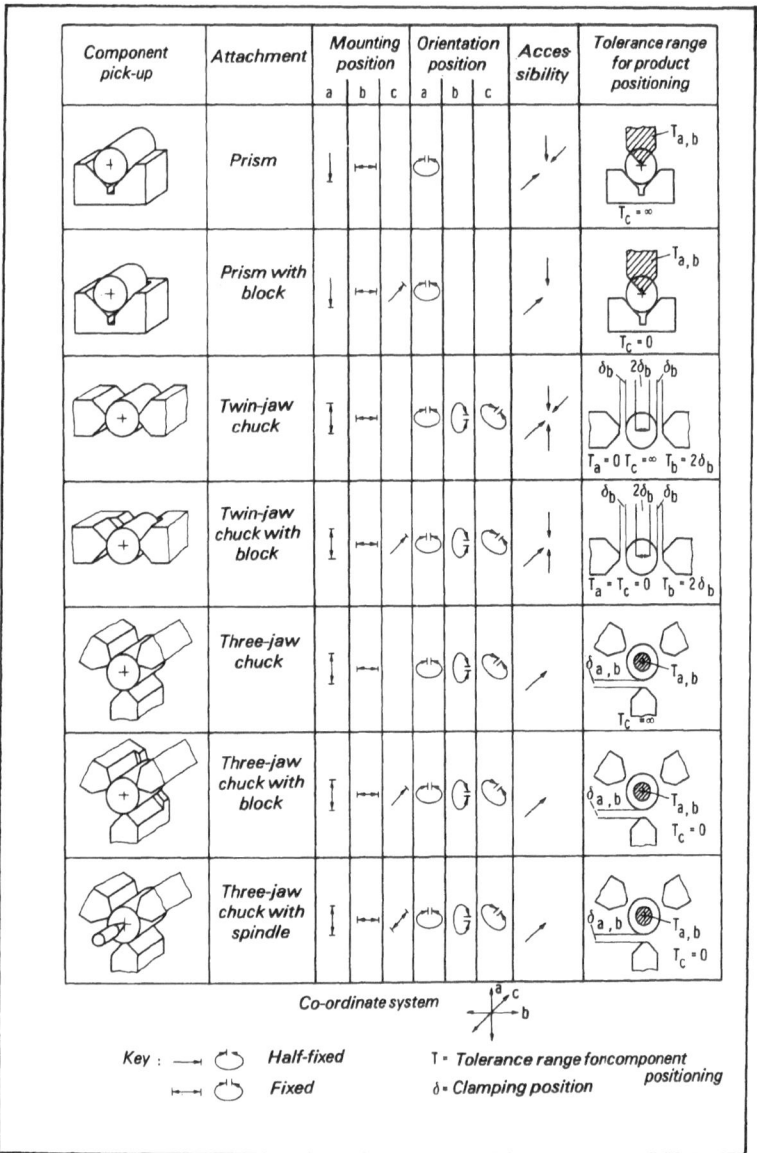

Fig. 10: Handling features of component pick-ups for symmetrically rotational parts

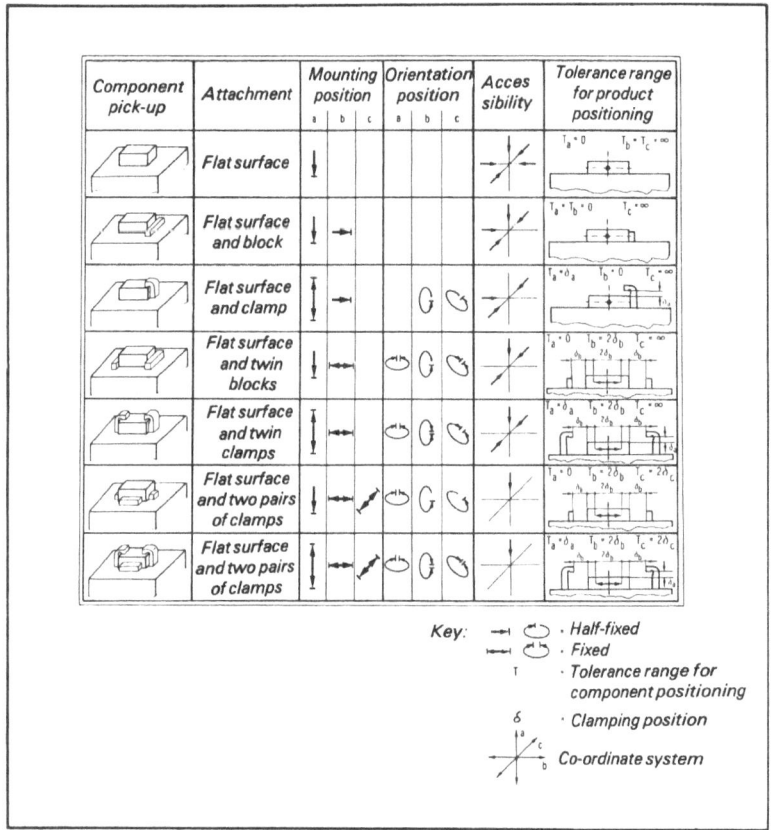

Component pick-up	Attachment	Mounting position a	b	c	Orientation position a	b	c	Accessibility	Tolerance range for product positioning
	Flat surface								
	Flat surface and block								
	Flat surface and clamp								
	Flat surface and twin blocks								
	Flat surface and twin clamps								
	Flat surface and two pairs of clamps								
	Flat surface and two pairs of clamps								

Key: — Half-fixed
— Fixed
T · Tolerance range for component positioning
δ · Clamping position
· Co-ordinate system

Fig. 11: Handling features of component pick-ups for cubic components

Here accessibility is confined by the arrangement of the guides more than in the case of turning machines having horizontal beds. Apart from the rarely installed turning machines with suspended shafts, accessibility from above is also given.

Accessibility in the shaft direction can be increased by horizontally and vertically suspended shaft arrangements in which components are fed through the shaft, given restricted component diameter.

Next to production units themselves it is important to consider their relative position one to another. This is important for the transfer of components between machines since largely through this will the observed position of components be determined. This is particularly

Horizontal turning machine

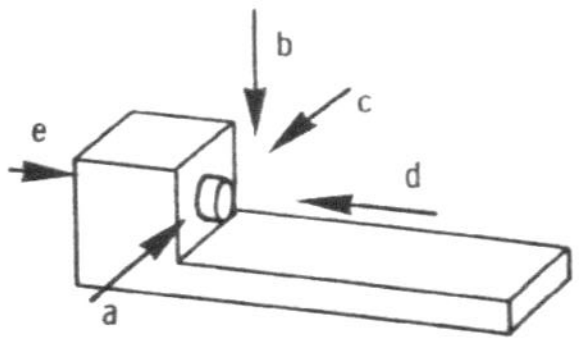

Vertical turning machine

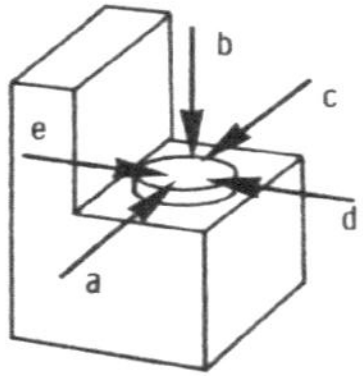

Machine construction		Accessibility from direction					Notes
Shaft direction	Construction of guides	a	b	c	d	e	
Horizontal — Rectangular machine bed	Long bed	2	2	2	1	0	Accessibility from e with aid of rods
	Short bed	2	2	2	1	0	
Sloping bed	Long bed	2	2	0	1	0	Accessibility from c with opening in machine bed / Accessibility from e with aid of rods
	Short bed	2	2	0	1	0	
	No bed	2	2	1	2	0	Accessibility from e with aid of rods
Vertical standing — Single position machine	External production	2	2	0	2	1	
	Internal production	2	0	0	0	2	
Dual position machine	With cross-beams	2	0	2	0	1	
	Without cross-beams	2	0	2	0	1	
Vertical suspended		2	0	2	0	0	Accessibility from b with aid of rods
Sloping		1	1	2	1	2	

Fig. 12: Accessibility of single shaft turning machines

Key:
0 = No accessibility
1 = Restricted accessibility through additional apparatus
2 = Accessibility given

relevant for linked manufacturing facilities in which transfer between several machines directly through handling equipment without transfer onto conveying systems should be possible. As in the example of turning machines shown (see Fig. 13), the number of necessary handling axes depends in reality on the relative positions of the shafts.

Just as with matters of space, so the time factors of production facilities affect materials handling. The time factors for storage, conveying and manufacturing facilities are determined by the necessary times for the performance of single production processes as well as the storage capability of these facilities, since this can make possible a decoupling of handling, storage, conveying and manufacturing processes[61-63].

As manufacturing facilities generally are not able to store components, the time necessary for manufacture determines the component handling time. The analysis of manufacturing times generally provides a wide spectrum of differing time claims of the machine tools. Some components have extremely long manufacturing times (e.g. in large-scale mechanical engineering) while others have extremely short manufacturing times (e.g. in small scale technology – up to approx. 0.1 seconds). Both extreme values present problems for automatic handling.

Extremely long manufacturing times can mean that handling equipment does not work very efficiently. A way out is offered by the possibility of including more machines. But these need to be serviced. However, the maintenance of several manufacturing facilities served by a single piece of handling apparatus can lead, at least in the short term, to the handling equipment being overloaded with the simultaneous change of components for several manufacturing facilities.

In the process of manufacturing components with extremely short manufacturing times, the possibility of automatic servicing is limited by the restricted movement speed of the handling equipment through its dynamic behaviour.

Closer consideration of the time factors of manufacturing facilities shows that the major times (the tool change times and the component change times) insofar as they are machine-dependent, affect the performance of handling processes.

Both the duration and the sequence of these time factors influence the maximum additional cycle time of a component handling system. Fig. 14 is a table showing the influence of sequence in insertion and removal of components from pick-up points. A distinction will be made between the following when applied to part pick-up:

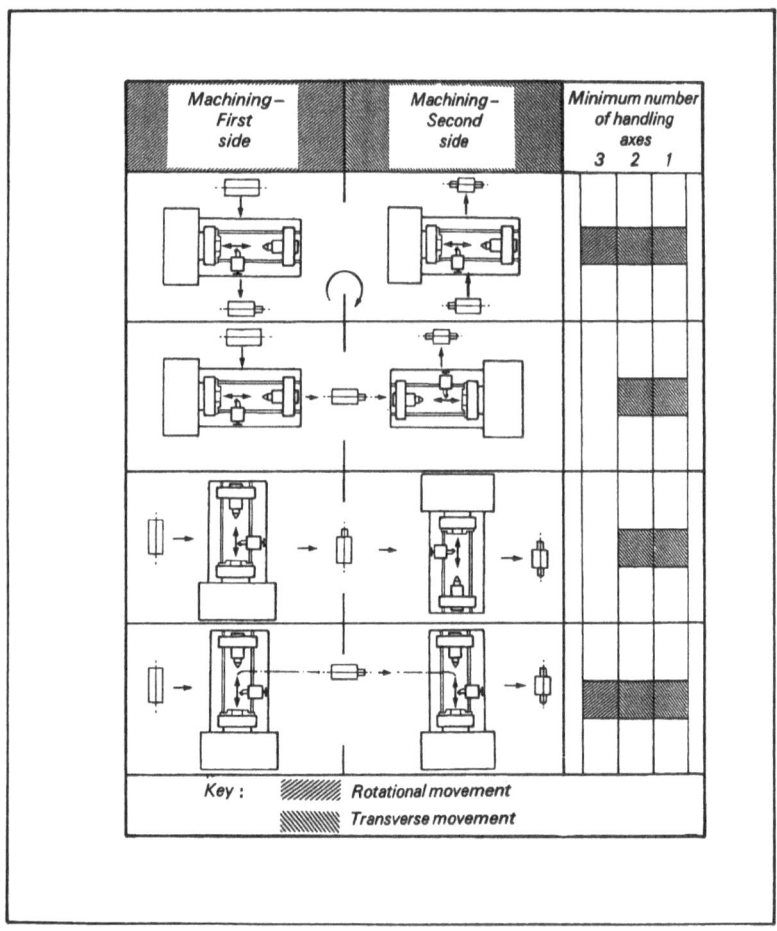

Fig. 13: Influence of machine installation on transferring processes on turning machines

- One or several components.
- In linear or circular arrangement.
- In equal or different clamping positions.

From the sequence of handling processes arise various possibilities for component movement and thus different cycle times for the handling apparatus.

The influence of sequence of production processes, with the

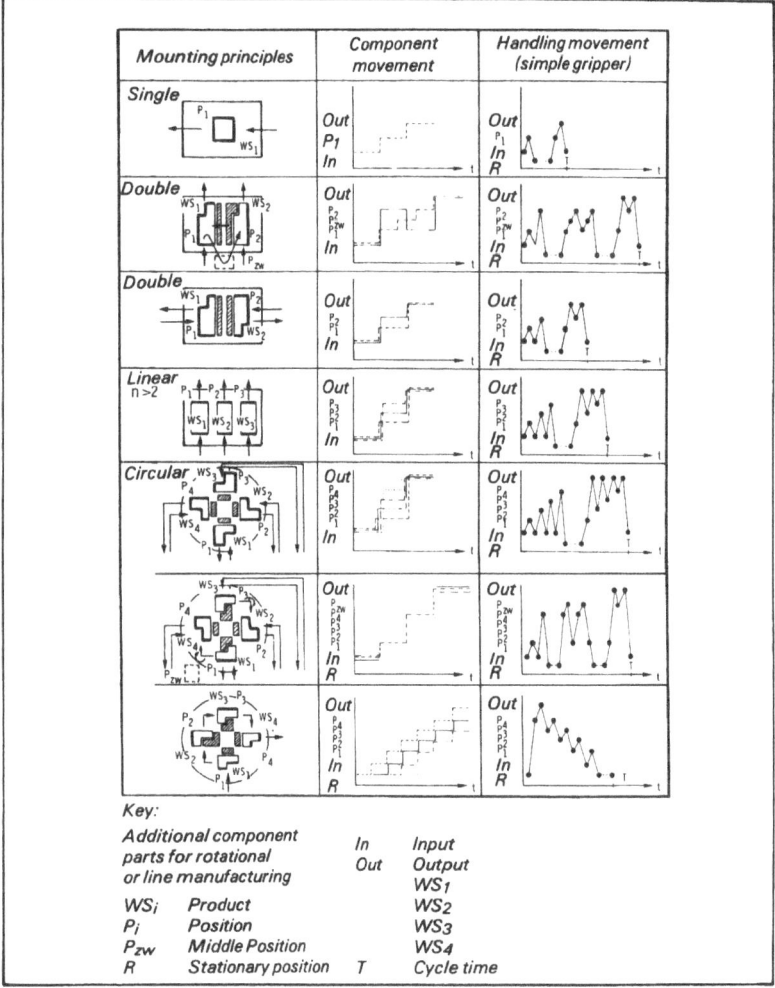

Fig. 14: Alternative movements depending on differing mounting principles in milling machines

simultaneous mounting of several components, is shown in Fig. 15.

From the various combinations of alternative arrangements for successive or simultaneous manufacturing, differing times for the performance of handling processes, both within and without machine workspaces, are produced.

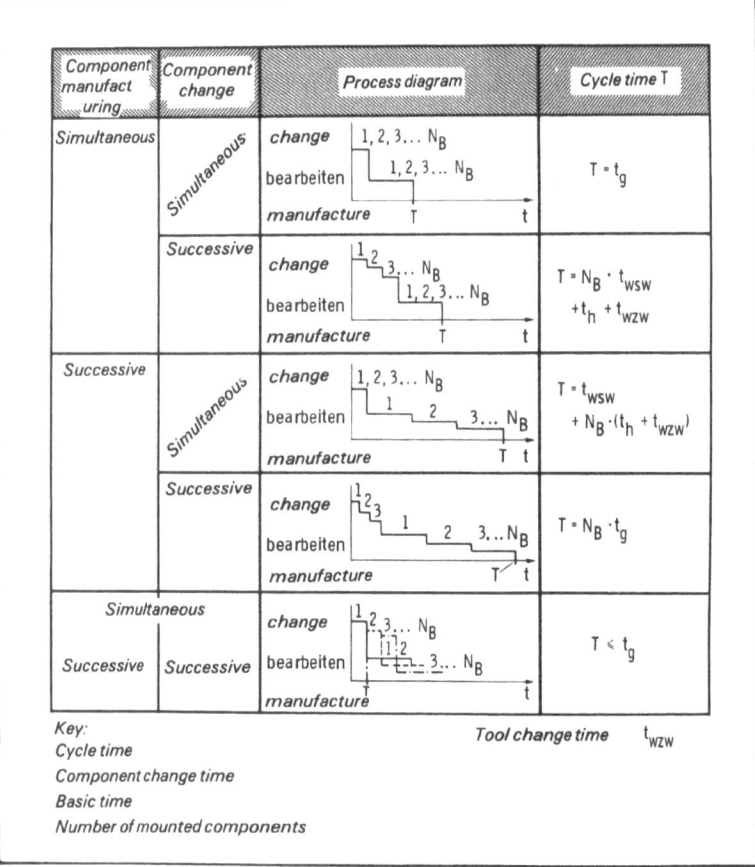

Component manufacturing	Component change	Process diagram	Cycle time T
Simultaneous	Simultaneous	change 1, 2, 3... N_B bearbeiten 1, 2, 3... N_B manufacture T t	$T = t_g$
	Successive	change 1 2 3... N_B bearbeiten 1, 2, 3... N_B manufacture T t	$T = N_B \cdot t_{wsw}$ $+ t_h + t_{wzw}$
Successive	Simultaneous	change 1, 2, 3... N_B bearbeiten 1 2 3... N_B manufacture T t	$T = t_{wsw}$ $+ N_B \cdot (t_h + t_{wzw})$
	Successive	change 1 2 3 bearbeiten 1 2 3... N_B manufacture T t	$T = N_B \cdot t_g$
Simultaneous Successive	Successive	change 1 2 3... N_B bearbeiten 1 2 3... N_B manufacture t	$T \leqslant t_g$

Key:
Cycle time   Tool change time t_{wzw}
Component change time
Basic time
Number of mounted components

Fig. 15: Dependence of cycle time on alternative possibilities of
manufacturing of several simultaneously mounted components

As well as a study into the space requirements and the time factors
involved in storage, conveying and manufacturing facilities, a know-
ledge of the processes occurring in the production system for the
integration of handling operations is necessary. To this end a functional
model to describe the production processes will be developed in the
following section.

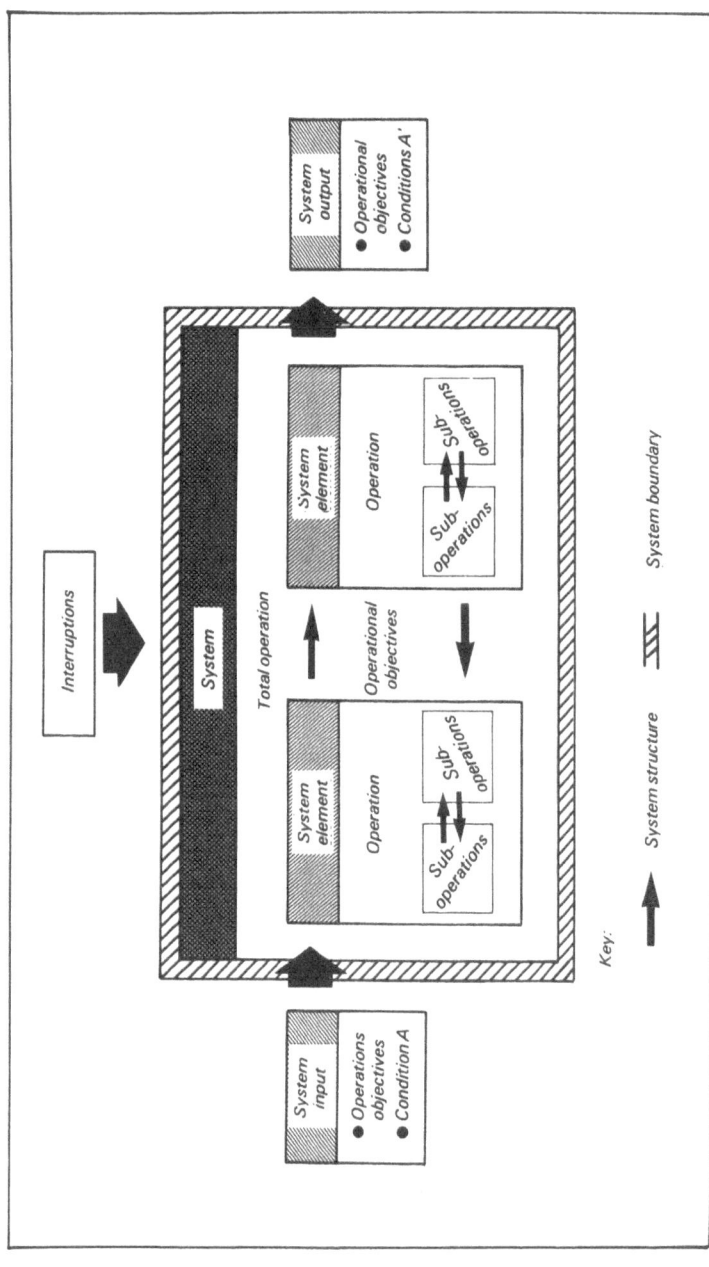

Fig. 16: System operations

Functional analysis of production systems

As a basis for integrating handling operations in production plants, the systematic analysis of the production system processes assumes in the first instance the definition of the terms 'system' and 'operation'. In the scope of this study, only technical systems are being considered, in contrast to, for example, a mathematical treatment (Fig. 16)[64].

A system is defined by its operations, objectives, operational units and the conditions of the operations objectives at the input and output of the system.

An operations objective is the purpose of carrying out an operation, for example, a component can be an operations objective, provided it is changed in the manufacturing process.

The function of a system is to vary the condition of an identified operations objective, that is, to alter defined features of it through the geometry of a component, or to preserve it within the system from the influence of external disturbances. An operation can affect several features of an operations objective. Thus, for example, not only can the shape be varied but the surface area, weight, dimensions and so on.

Operations can be further described by one or more characteristics (operation parameters). This is how movement through space, time, speed and so on are defined. A system is separated from other systems by its system boundary, which includes the elements of the system (operating units). The operating units (e.g. the machine bed, drive, tool clamp etc. of a turning machine) are related to each other within a system structure. In technical systems, the system structure reproduces the causal and time connections of the operating units.

The operating units together carry out the total operation of the system. The operating units themselves can be understood as independent systems with sub-operations. The operations objectives appropriate to this system are not necessarily the same objectives which are influenced by the total system. Thus the tooling in a turning machine is an operations objective of the support system, since it is moved by this, while the operations objective of the total 'machine tool' system is the component.

Operation complexes are sequences of operations. 'Process' is the general heading for operations complexes, operations and sub-operations, etc.

Definition of functional model. With the help of the above definitions under the heading of 'system' it is possible to construct an operational model to describe the processes within production systems. The representation of operating processes does not have to be linked to the existence of operating units. It is more useful for the development of new production facilities to consider operating processes chiefly

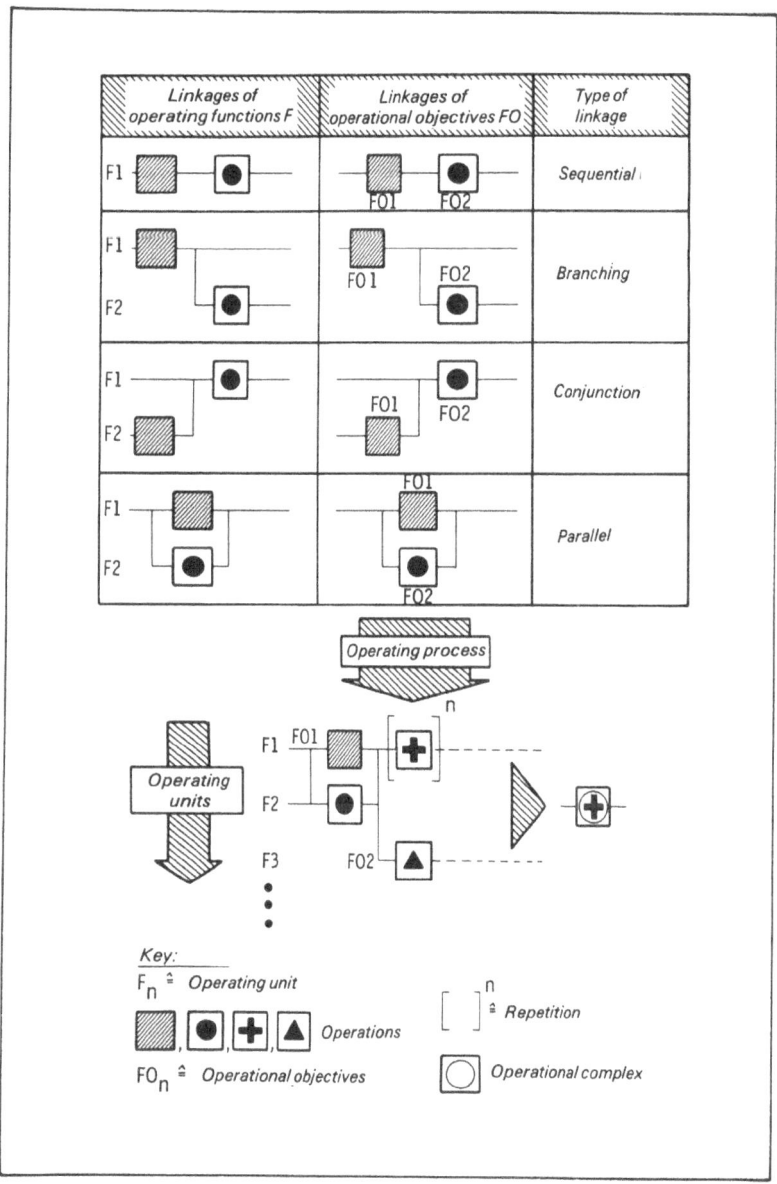

Fig. 17: Construction of possible operations procedures

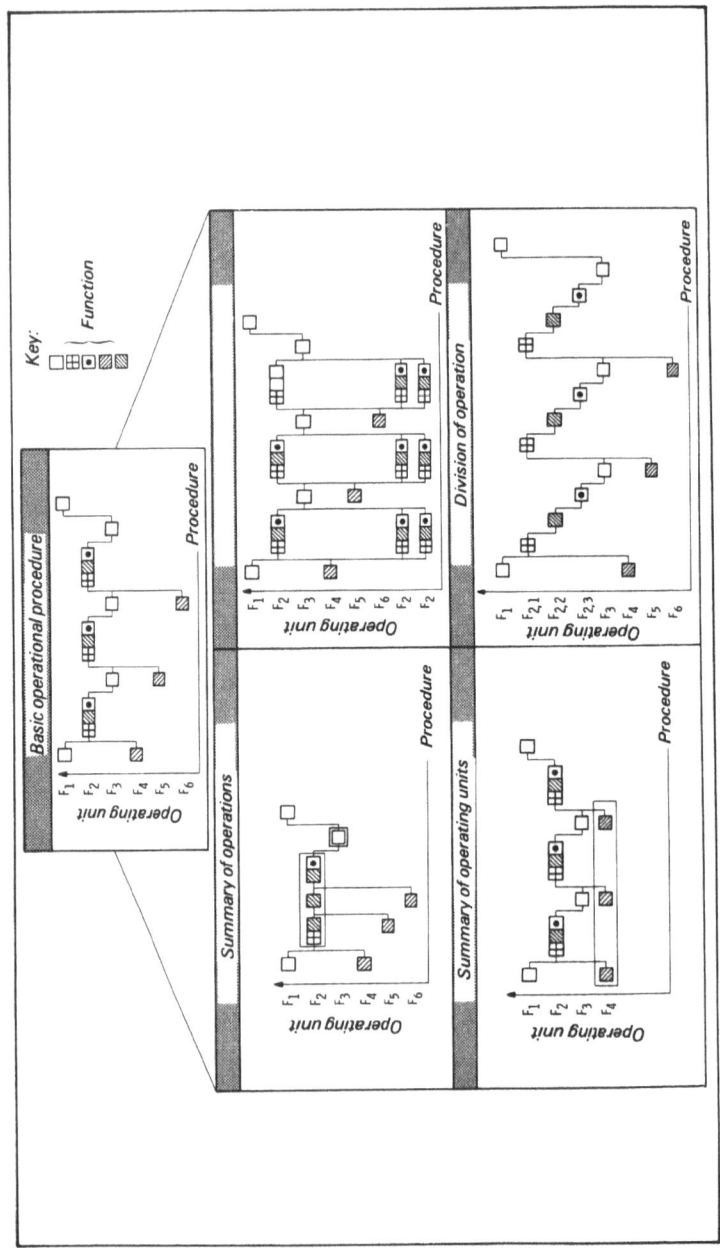

Fig. 18: Principal alternatives for manipulating operations procedures

disassociated from existing technical solutions[65,66].

In order to create operations procedures (operation networks), the possible link-ups of operating units and operation objectives will be investigated[67]. Fig. 17 shows the alternatives.

In principle, operations procedures can link differing operations, operating units and objectives together. This can happen sequentially where an operating unit carries out various operations in logical and chronological order. This allows the possibility of following one operation on one objective with another operation on another objective.

As an example, mention should be made of flexible handling equipment which is capable of extending from the handling of a component to the deburring of another component with the help of tooling, i.e. another operations objective.

At branching points of operations procedures, the execution of one operation by a specific operating unit follows an operation by different operating units. The execution of the second operation depends on the completion of the first. In contrast, the operating unit, after carrying out the first function, can undertake optional operations after the branching point. The conjunction is the exact reverse of the branching linkage.

A combination of branching and conjunction linkages creates a parallel linkage. The parallel execution of several operations does not necessarily require these operations to be simultaneous.

These operation linkages can be used to produce complex operations procedures. In this context reference can be made to operations complexes if the operations procedures occur within enclosed processes. Operations or operating sequences can be repeated in operations complexes. A new concept of operations is usually defined for such complexes.

To achieve the proposed configuration of operations procedures, it is worth changing the sequence of operations, working from a basic procedure. The possible outcomes are presented in Fig. 18.

In the first instance, repeated operations procedures can be combined. For example, this is the case with the transition from a machining process which must be changed several times to a comprehensive machining process. Here the 'machining' operation can be combined. In this case the capacity of the 'machine tool' operating unit must be increased by the use of the appropriate additional equipment, such as turntable, pivoted table, pivot and so on.

In the same way similar operating units can be brought together, if, for example, the handling of materials by several separate movement mechanisms is replaced by complex handling apparatus.

Where operations objectives are brought together, similar operations are carried out on different objectives. An example of this would be multiple gripping equipment which can pick up several components simultaneously.

The division of operations is where an operation or an operational complex is taken over by several operating units.

Operations of handling facilities. With the help of functional models, handling systems can be defined together with an analysis of the processes within these systems (Fig. 19).

The total operation is the handling of the objectives concerned. These can be components or component carriers such as pallets or containers (Fig. 20). Tools, attachments, measuring devices, lubricants and waste can also count as objectives for handling. These, however, are not considered in this study.

Components can be handled as single items or in defined quantities, such as a load. It is essential for the handling of component loads that the loading facility is determined by the geometry of the component and the condition of the surface area. In estimating alternative methods for the automatic handling of component loads, account should be taken not only of handling the load as a whole, but of handling single items which go to make up the load and when it is separated. Experience shows that in particular the separation of individual components is often hindered by their features, such as a tendency to catch, stick or jam. The advantage of one load lies in a reduction in the number of handling processes, since all single parts of a load can be handled together.

Component carriers are set up not only in storage and handling units but in manufacturing facilities. While the handling conditions in component or load handling will vary according to the different characteristics of the components (such as shape, dimensions, weight, surface area, and so on), certain wide-ranging standard conditions are necessary for component carriers.

As with the handling of component loads, care has to be taken with component carriers so that additional handling processes do not cause problems through the loading or unloading of the carrier. There is a difference here between pallets and containers. With pallets, special structural features define the position, orientation and quantity of individual articles to be carried. However with containers, only the maximum volume capacity for individual parts is given. Orientation, position and quantity can vary within this volume.

The nature of the objectives to be handled is determined at the system input and output by the characteristic position (place), orientation (rotational position), quantity, order and point of time during the handling process (time sequence). Operations can be defined for a handling system which will produce the desired changes in these component characteristics.

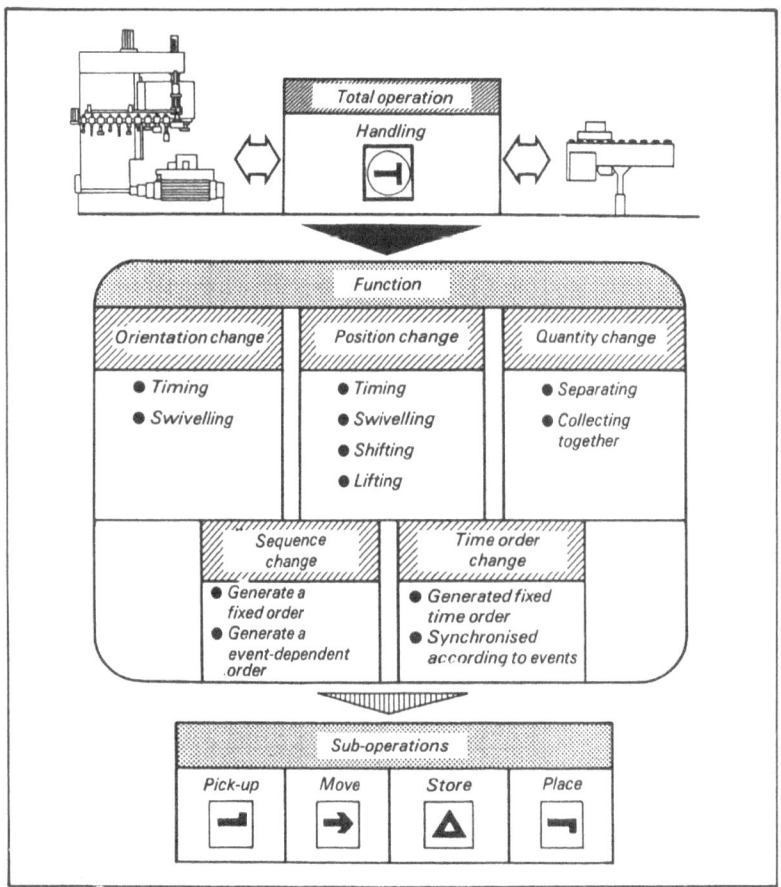

Fig. 19: Functional analysis of a handling system

Handling operations can be determined more clearly with the aid of alternative methods of carrying out these operations. An orientation change can be achieved only by turning or swivelling the object being handled. Here a distinction must be made between the processes to generate a load (unordered components), the processes to generate a load with components in some preferred and defined orientation, and the processes required to generate components in a single defined orientation.

Of course there can be change of quantity when components are separated or taken away from a group of other components. This can

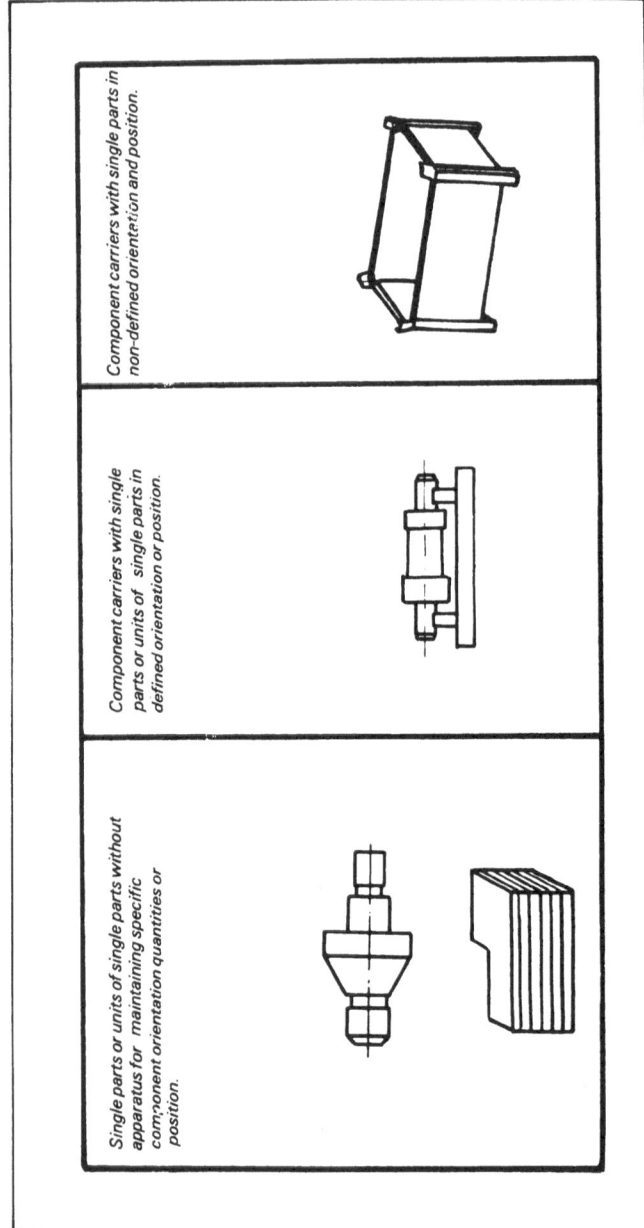

Fig. 20: Objectives to be handled

happen both with an unknown quantity of components and with an exactly defined quantity of components. The collecting together of components means the reverse of these processes. Components which are defined with regard to position, orientation and quantity are referred to as 'ordered' components.

The need to change the order of components arises, for example, when several different components, which will later be assembled together as a finished product, must be manufactured together for convenience. One example of this is the split bearing housing for pistons. Here the required raw materials for the two halves of the bearing are delivered in large quantities in containers to the machining unit. In this case, handling means, among other things, that at any given time an upper and a lower bearing housing will be presented alternatively to the machining unit. This generates a fixed sequence which will be maintained through to final assembly.

The sequence can be changed according to events. For example, after an inspection process sub-standard parts are identified and they must be separated out from the material flow before the next station.

In all cases, when the storage, conveying and manufacturing facilities which require a handling process demand a differing time behaviour, the timing of the handling processes will be affected. According to the inflexibility with which fixed time intervals must be observed, a re-timing (rigid timing) or a synchronisation (event-flexible timing) can be used.

Changes in the orientation and position of a component require any orientation to be achieved by technical means which can be achieved by any turning, swivelling, shifting and lifting equipment. In contrast, changes of quantities, sequence and time order tend to happen under organisational direction. Even so, the carrying out of these operations still demands technical assumptions to be made, requiring equipment to be put in a position that allows the material flow to be led off, for example, by a switching device, or to put components into temporary store. The execution of these operations always requires orientation and/or position changes.

It is essential that all handling operations can be broken down into the sub-operations of 'pick-up, move, store and place'. Under the sub-operation 'pick-up' are included all the processes which serve to recognise the orientation, position, quantity, sequence and presence of components, as well as the process of bringing the necessary power to bear on to the component to achieve the next sub-operation (e.g. gripping).

In contrast, these functions remain constant during the sub-operation of 'storage'. Storage serves to maintain the condition of a component so that it is available at a particular time. Placement comes at the end of

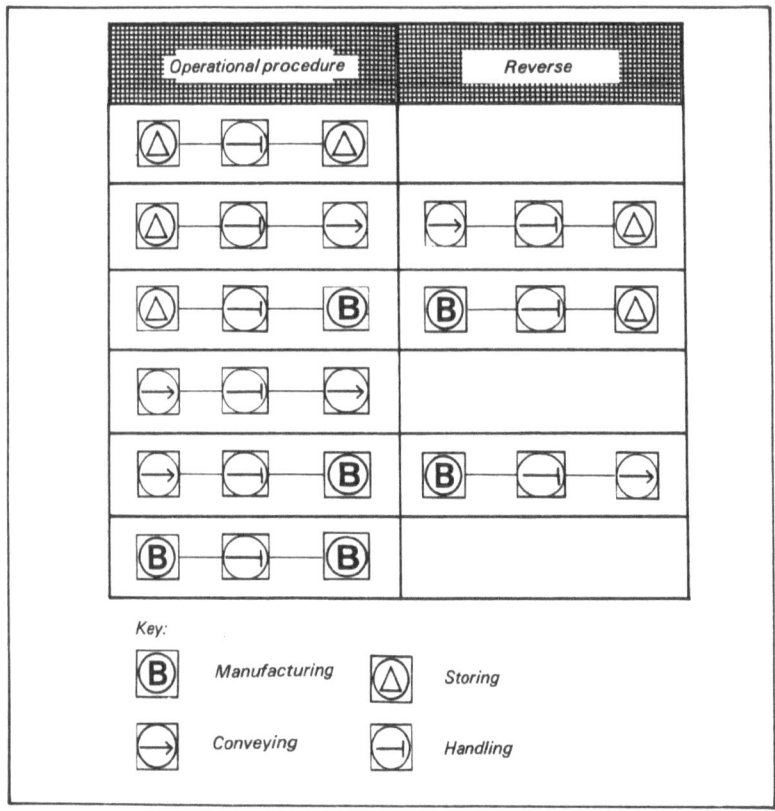

Fig. 21: Principal alternatives for the arrangement of handling processes as connecting links between storage, conveying and manufacturing facilities

the handling process, at which point all power required for handling is removed from the component. Thus placing can be seen as a complementary operation to pick-up. Pick-up and place are often therefore carried out by the same operating units.

To transfer components between manufacturing, conveying and storage facilities, that is the carrying out of the operations complex 'handling', a repeated run-through of handling operations is often necessary. Thus handling processes among all production plants involving conveying, storage and manufacturing can be undertaken (Fig. 21).

Handling operations from VDI 2860 (outline)		Handling sub-operations			
		Pick-up	Move	Store	Place
Storing	Storing, ordered	●	●	●	●
	Storing, partly ordered	●	●	●	●
Quantity change	Divide	●	●		●
	Separate	●	●		●
	Separate into lots	●	●		●
	Allocate	●	●		●
	Branch off		●		
	Collect together		●		
Moving	Turn		●		
	Swivel		●		
	Swift		●		
	Orientate	●	●		●
	Position	●	●		●
	Order	●	●	●	●
	Pass on	●			●
Securing	Hold	●			
	Clamp	●			
	Loosen				●
Controlling	Test	●	●	●	●
	Measure	●	●	●	●

Fig. 22: Comparison of handling sub-operations with the operations terms used in 'VDI 2860'

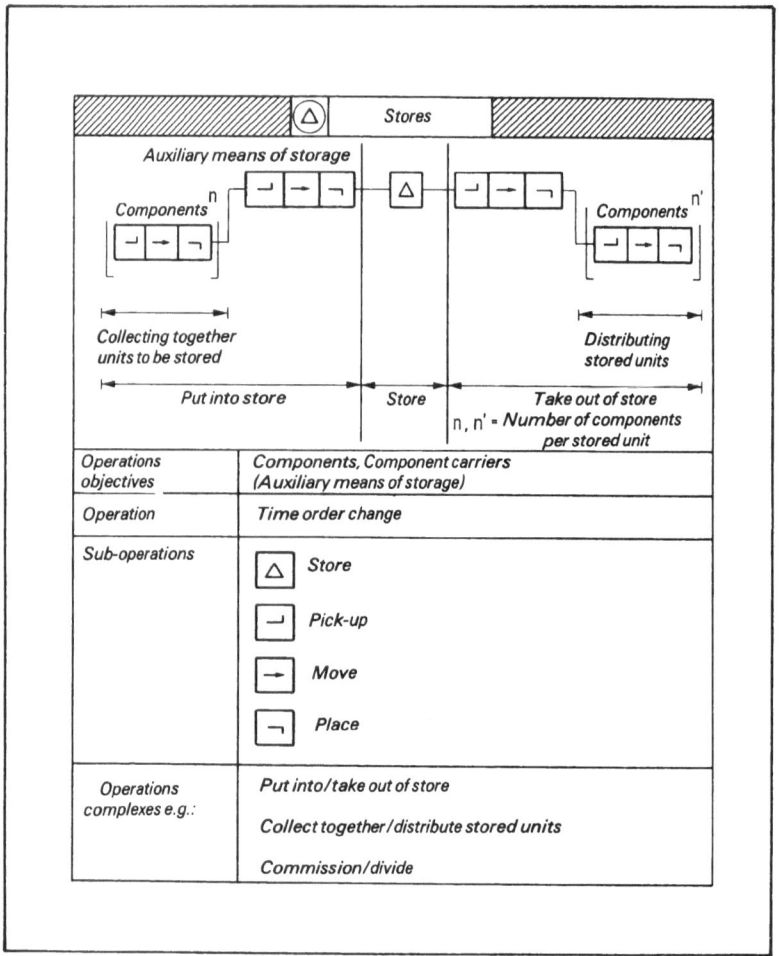

Fig. 23: Functional analysis of storage systems

The proposed definition of handling functions, which departs from the existing classification, was undertaken with the objective of delimiting individual operations which were clearly distinct from one another[68]. This has achieved a definition of processes which is both uniform and reproducible, leaving the requirements on individual operations to be equally distinctly determined. The operations classifications used by the VDI (Association of German Engineers) can be

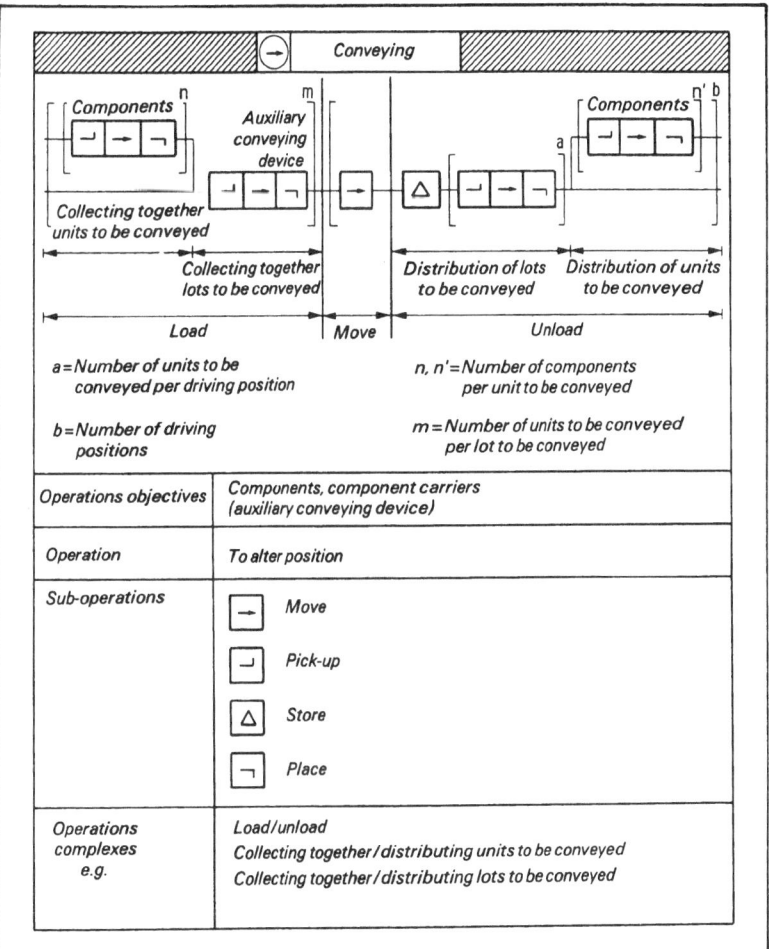

Fig. 24: Functional analysis of conveying systems

explained by a combination of the terms used here. The VDI handling operations can thus be explained as operations complexes of the sub-operations 'pick-up, move, place and store' (Fig. 22).

One exception is the VDI handling operation 'controlling'. According to the definitions used in this study, controlling is an independent process within which handling operations can be included.

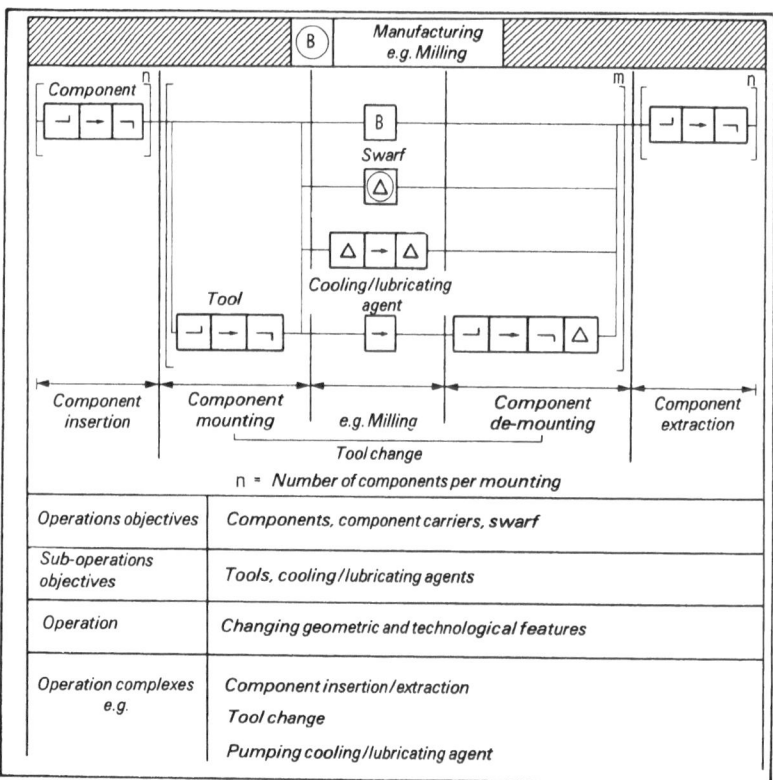

Fig. 25: Functional analysis of manufacturing systems

Operations of storage, conveying and manufacturing facilities. With the aid of the functional model, the processes within storage, conveying and manufacturing facilities can be analysed. No fixed operational procedures can be defined in this way. Rather, it is clear that as in the linkage alternatives in Fig. 18, a variety of different operational procedures can be represented.

The operations objectives of a store are components or component carriers in the form of auxiliary means of storage, such as pallets and containers. The function of a store is to maintain the condition of a stored object – in particular its characteristics relevant to the handling process, such as position, orientation, quantity, sequence and to make it available at a determined point in time. That is, that by means of a store, the time for starting a particular process can be determined (Fig. 23).

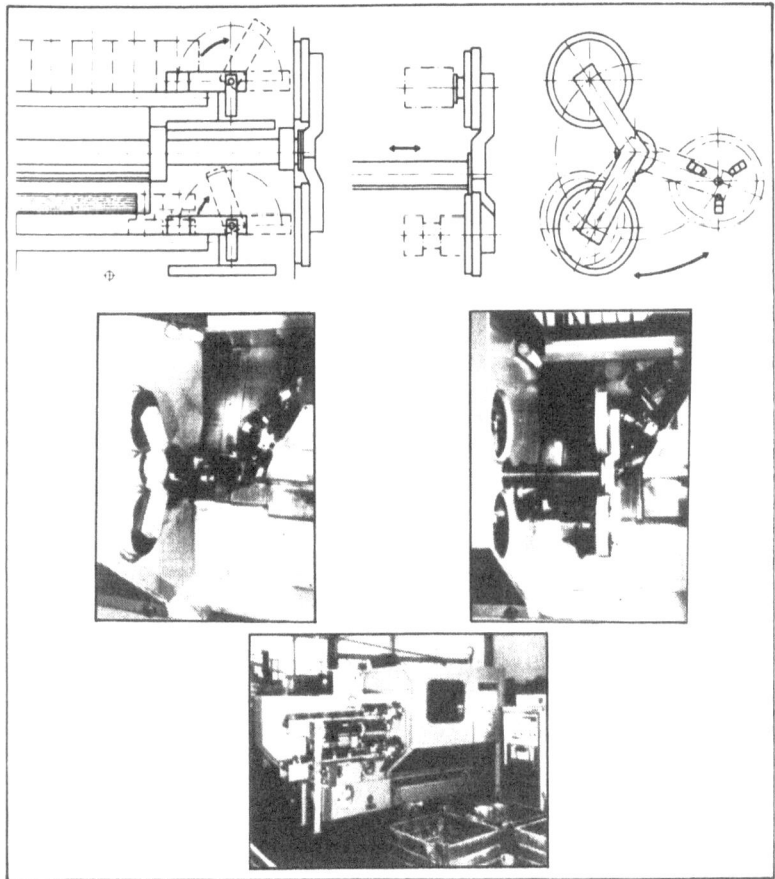

Fig. 26: Turning machine with component change apparatus

The operations objectives of a conveying facility are also components or component carriers (auxiliary means of conveying). The operation of a conveying system is to change the position of a component. Thus the orientation, quantity, sequence and point in time of arrival of an object to be conveyed are changed to some degree (Fig. 24).

While with conveying and storage systems only components and component carriers are operations objectives, in the case of the manufacturing operation a multiplicity of operations objectives are involved; these flow over the system boundary of the manufacturing system (Fig. 25).

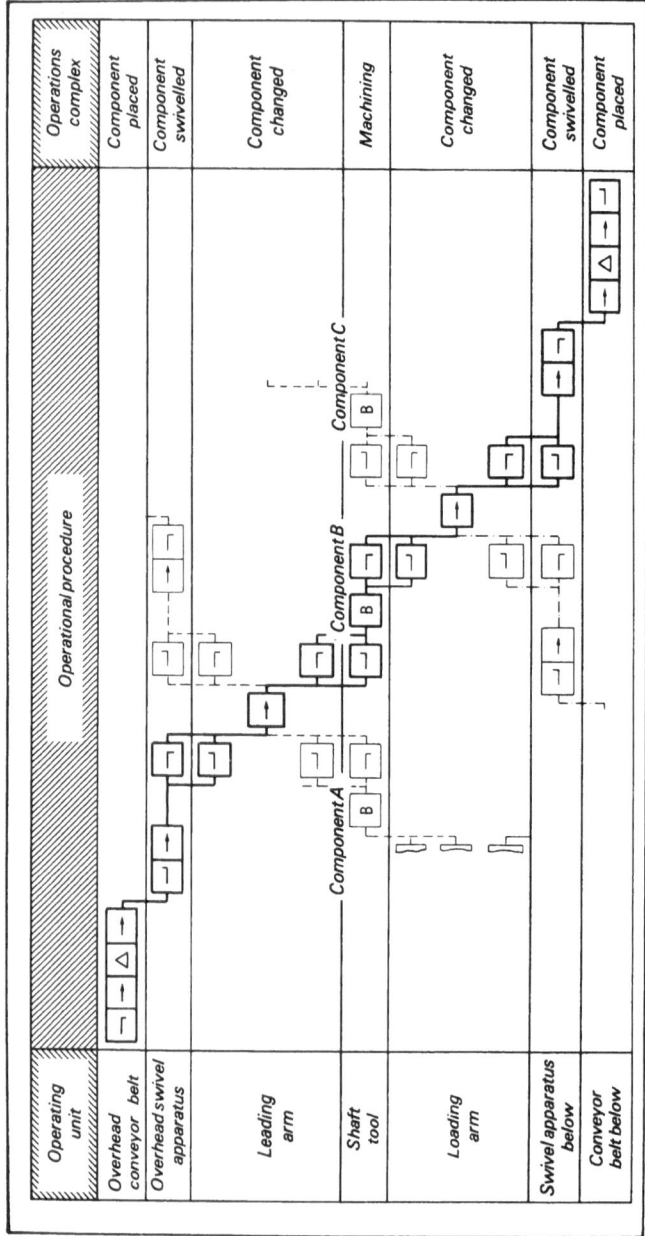

Fig. 27: Example of operational procedure in a manufacturing facility

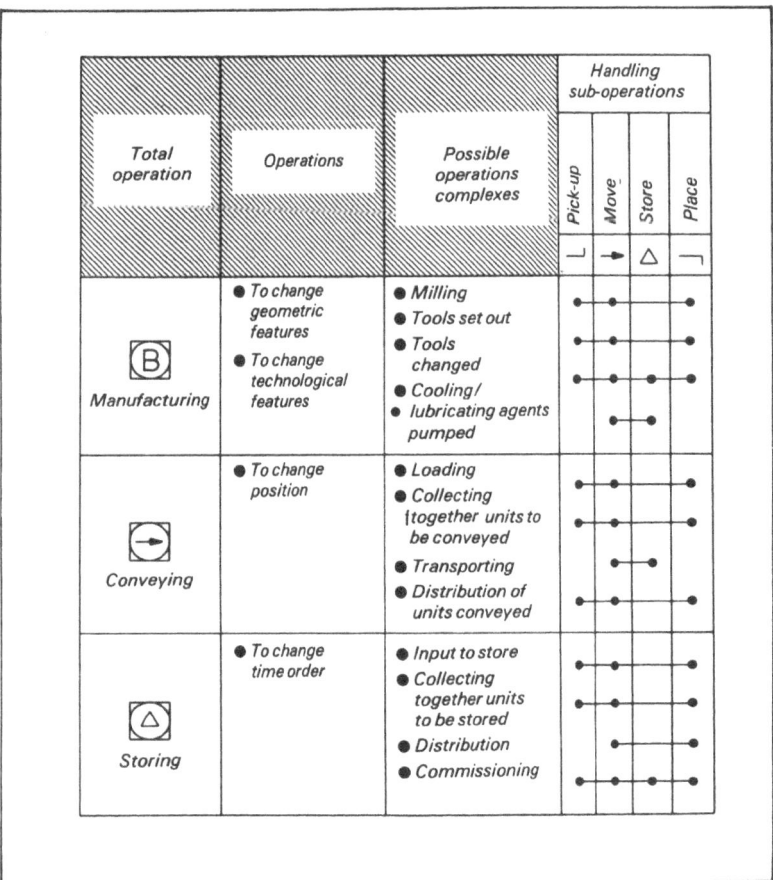

Total operation	Operations	Possible operations complexes	Handling sub-operations			
			Pick-up ⌐	*Move* →	*Store* △	*Place* ⌐
 Manufacturing	● To change geometric features ● To change technological features	● Milling ● Tools set out ● Tools changed ● Cooling/ lubricating agents pumped				
 Conveying	● To change position	● Loading ● Collecting {together units to be conveyed ● Transporting ● Distribution of units conveyed				
 Storing	● To change time order	● Input to store ● Collecting together units to be stored ● Distribution ● Commissioning				

Fig. 28: Handling sub-operations in comparison with possible operations complexes in storage, conveying and manufacturing systems

In the first instance those components whose geometric and technical characteristics will be changed during the manufacturing process must be identified. In this process, waste will be created – chips, swarf, shavings and so on – and these must be removed from the workspace. To carry out the manufacturing process, additional attachments, tools and cooling and lubricating agents will be required. These factors however are usually not deliberately changed in their characteristics.

The basis for the number of operations objectives in manufacturing systems in comparison with storage and conveying systems lies in

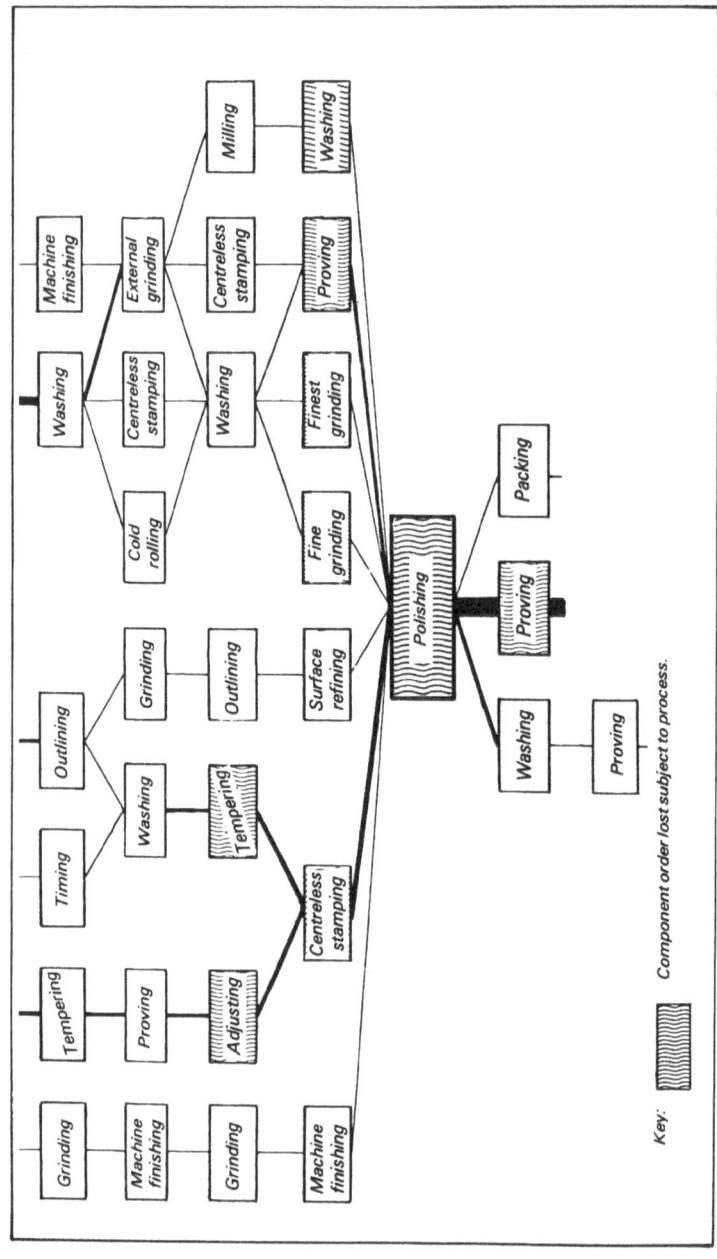

Fig. 29: Manufacturing operations of a production process for a range of components in precision engineering

complex operational processes. This is particularly the case if, to increase the capacity of the machine, additional operations are included; Fig. 26 shows an example of this[69].

In the picture is a turning machine for disc-shaped components in which the components are changed with the aid of two separate conveyor belts, a swivel apparatus and a loading arm having a dual gripper. The appropriate operational procedure is represented in Fig. 27.

In the representational figure, the repetition of operational procedures for tools, attachments, inspection equipment and swarf is excluded for reasons of clarity.

As the example and the analysis of storage, conveying and manufacturing facilities show, there is a great similarity between operational procedures in materials handling and processes in storage, conveying and manufacturing facilities. Fig. 28 states these assertions.

An investigation has been made of the degree of overlap between handling operations and operations complexes in production facilities. A complete overlap of operations concepts is shown. This facilitates, by the integration of handling functions in production plants, the use of sub-operating units of the manufacturing, storage, and conveying facilities to carry out handling function.

In the production process these types of operational procedures will necessarily occur in certain sequences. These will be linked by the manufacturing schedule to complex operations procedures.

These, in turn, can generally be presented in the form of a sequence of operations which can, however, depending on the complexity of the production process, seem very extensive and involved.

These complex operational procedures are mainly associated with a single component. However, in production it is usual that a variety of components have to be considered. Also it is common for the different components not to appear according to a technically correct and compulsory order. Much more often, the component sequence, and thus the sequences of operations to be carried out, is determined principally by the firm's current order situation. Thus constantly the operational procedures within a production system have to be altered so that they are no longer manually representable graphically.

Nevertheless, to achieve an overview of the operations in the production process, a diagram of manufacturing processes, as in Fig. 29, can be useful[70,71].

The diagram is an extract from the production procedures for a component range having a range of manufacturing operations. These manufacturing operations are derived from actual work schedules. In addition, the materials flow linkages of the manufacturing facilities are presented. With the materials flow direction which is derived from the sequence of manufacturing operations in the work schedule, an indication of the scale of materials flow is given. This in turn results from the numbers of components to be produced. The diagram can be expanded for studies of materials handling by the addition of technical handling information, such as weight and dimensions.

Principal methods of integration of handling operations

On the basis of the model developed within the functional analysis of storage, conveying and manufacturing facilities, operating units for the integration of handling operations in these facilities can be derived (Fig. 30).

To this end, the major operational procedures of materials handling in storage, conveying and manufacturing facilities must first be established, and these handling operations must be systematically divided between the individual operating units.

Following this, alternative solutions to carry out individual or combinations of handling operations for the operating units can be established. These alternatives are to be explained in detail in a further stage of the study.

The integration of handling operations means first of all the division of such operations between storage, conveying and manufacturing facilities (Fig. 31).

To put this into practice, the operating units in production facilities or additional handling facilities must undertake the handling sub-operations.

Integration of handling operations in storage facilities. First, the possible solutions for the performance of the 'pick-and-place' handling sub-operation can be studied separately from the possibility of integrating other sub-operations.

As already mentioned, this study is concerned with those operations which, in general, are performed by similar items of equipment, insofar as neither pick-up nor placing is assisted by pushers or ejectors. The performance of the pick-and-place operation is largely independent of the operating unit in which it is integrated. The various alternatives here can be distinguished according to the type of power input and the positioning of the component (Fig. 32).

Where the component pick-up is sited at one point, the position of a component is defined in terms of the mounting point. In contrast to this, the orientation of the component is optional. In general, only single components are handled with this kind of component pick-up, so there is no possibility of a quantity or sequence change. Because of the undefined orientation, this kind of component pick-up is used only in materials handling in exceptional cases.

Where component pick-up is part of a line configuration, component

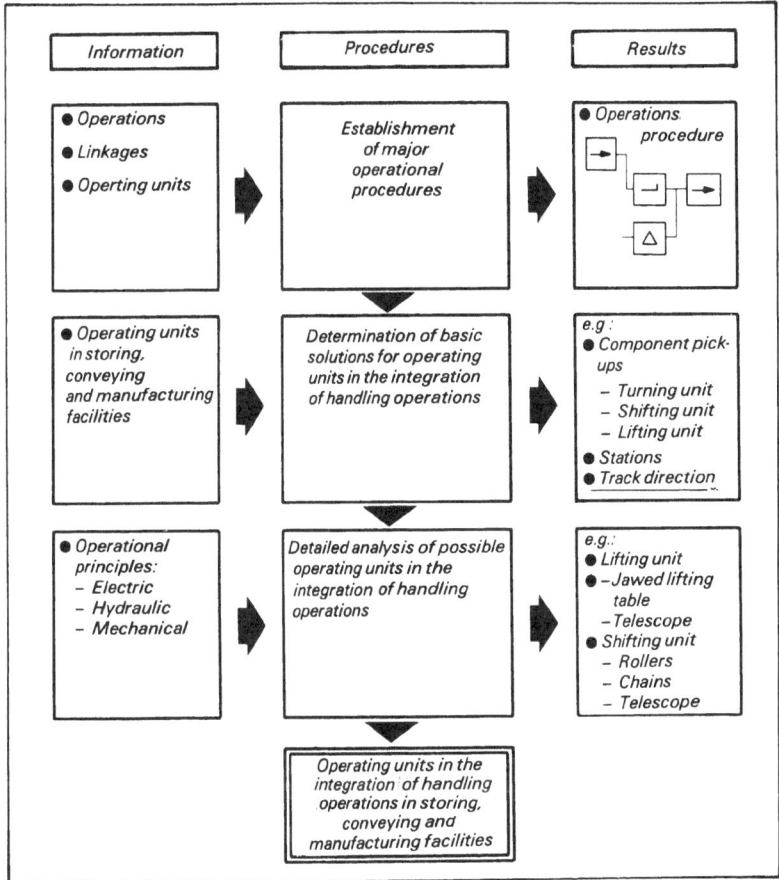

Fig. 30: *Procedures to determine alternative solutions for operating units in the integration of handling operations*

position along the mounting line is optional, provided it is not restricted by positioning elements. Thus, compared to a point configuration, the orientation is more precisely defined. With appropriate sizing, several components can be dealt with by a line configuration component pick-up. However if this configuration is used with a cotter, care must be taken that the 'place' sequence is determined from the 'pick-up' sequence.

The position of a component in a two-dimensional component pick-up is optional on the working surface, provided again that it is not

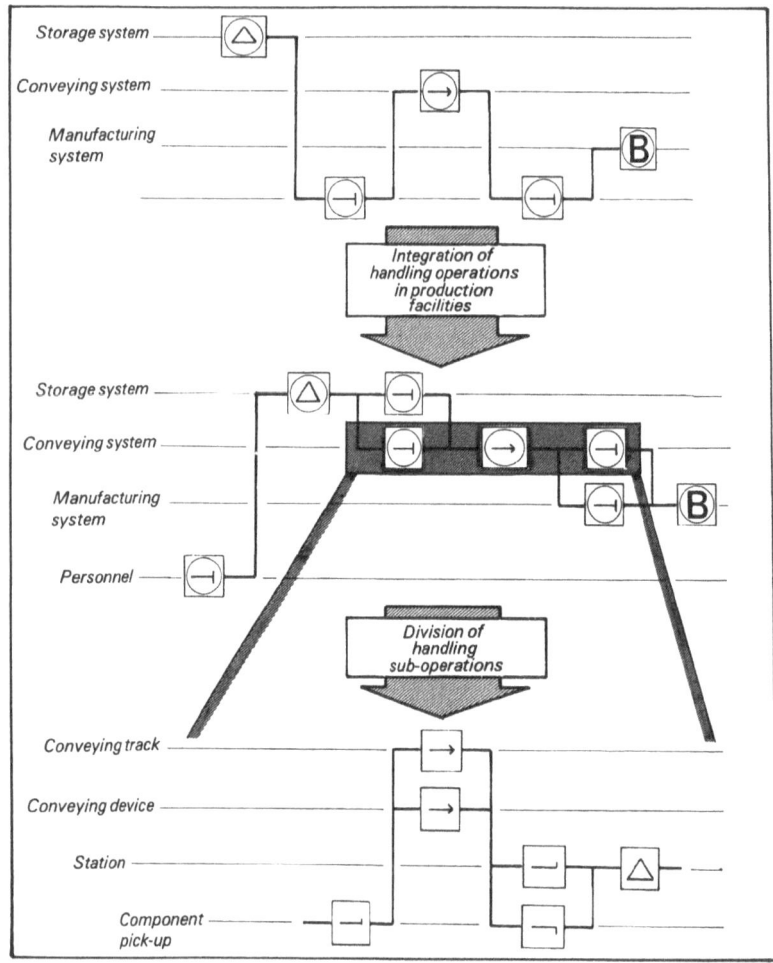

Fig. 31: Effect of integration of handling operations on operational procedures

restricted by positioning elements. The orientation of the component is fixed in one axis, although it may be further restricted by clamping elements. Two-dimensional component pick-ups can, if necessary, pick up several components. For three-dimensional component pick-ups, the above also applies.

With all component pick-ups, which can take up more than one component simultaneously, it must be remembered that the compo-

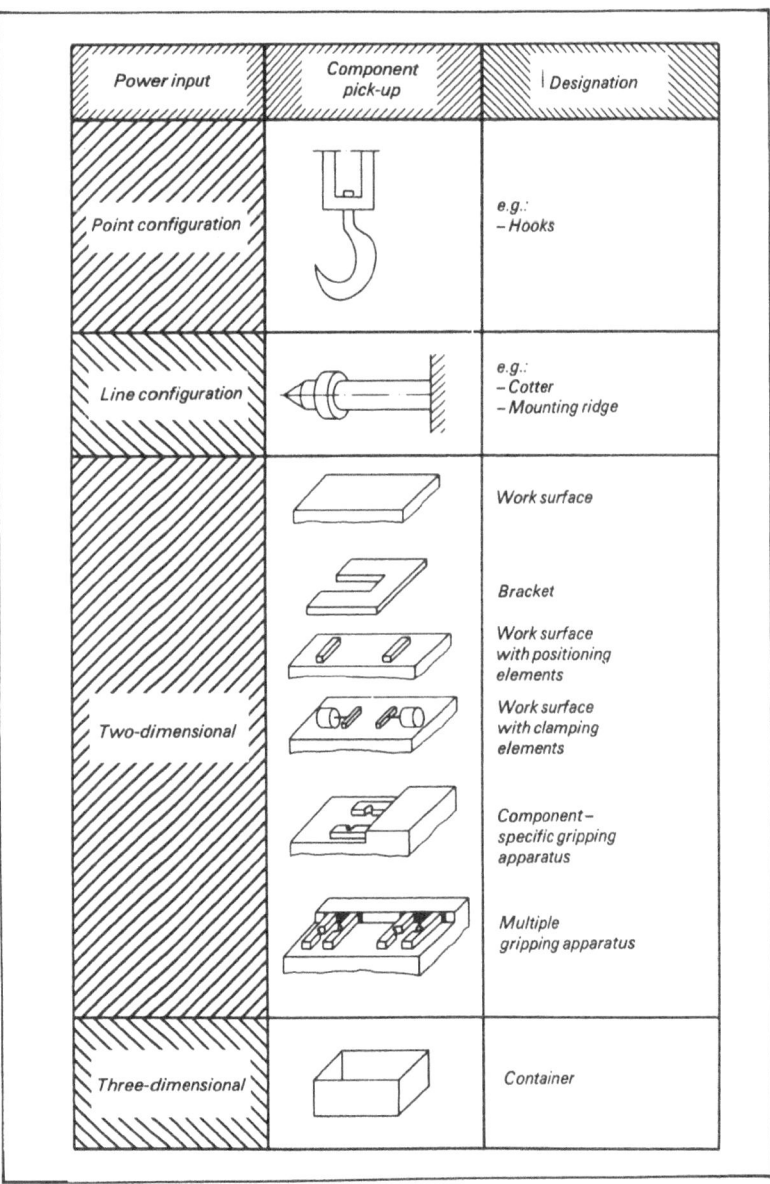

Fig. 32: Methods of performing the pick-and-place handling sub-operation

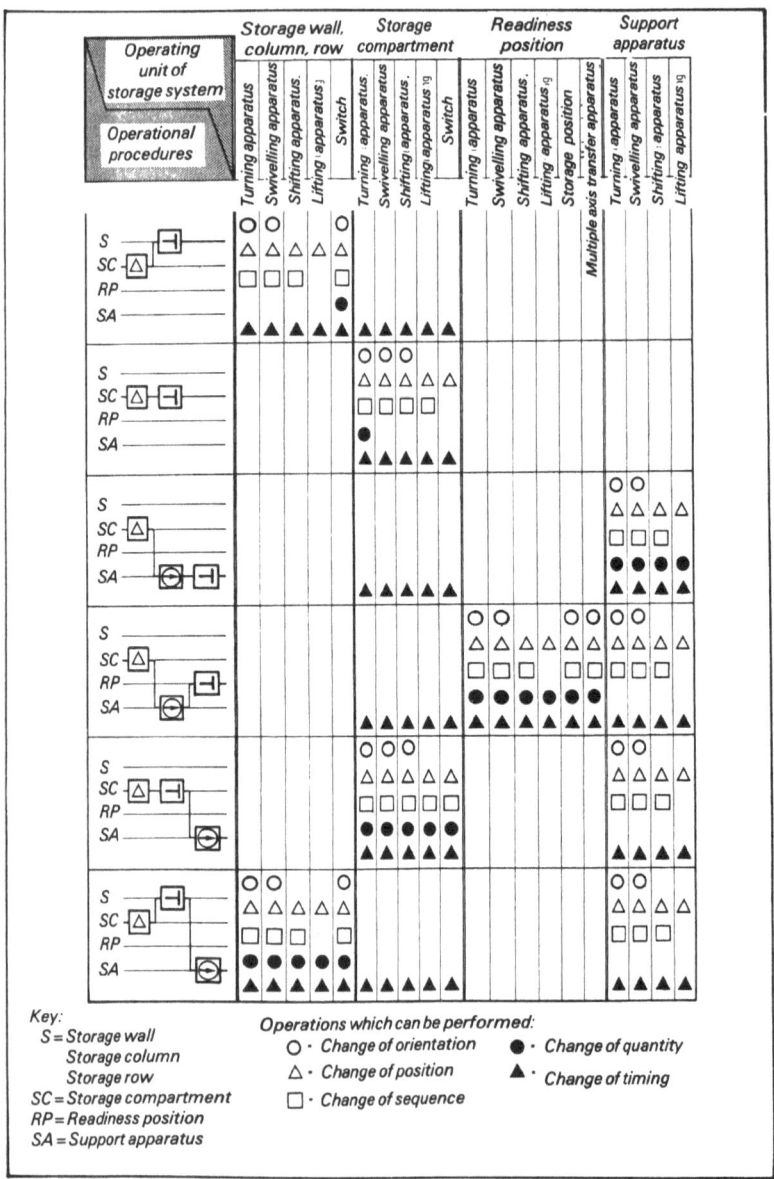

Fig. 33: Alternatives for integration of handling operations in the operating units of storage systems

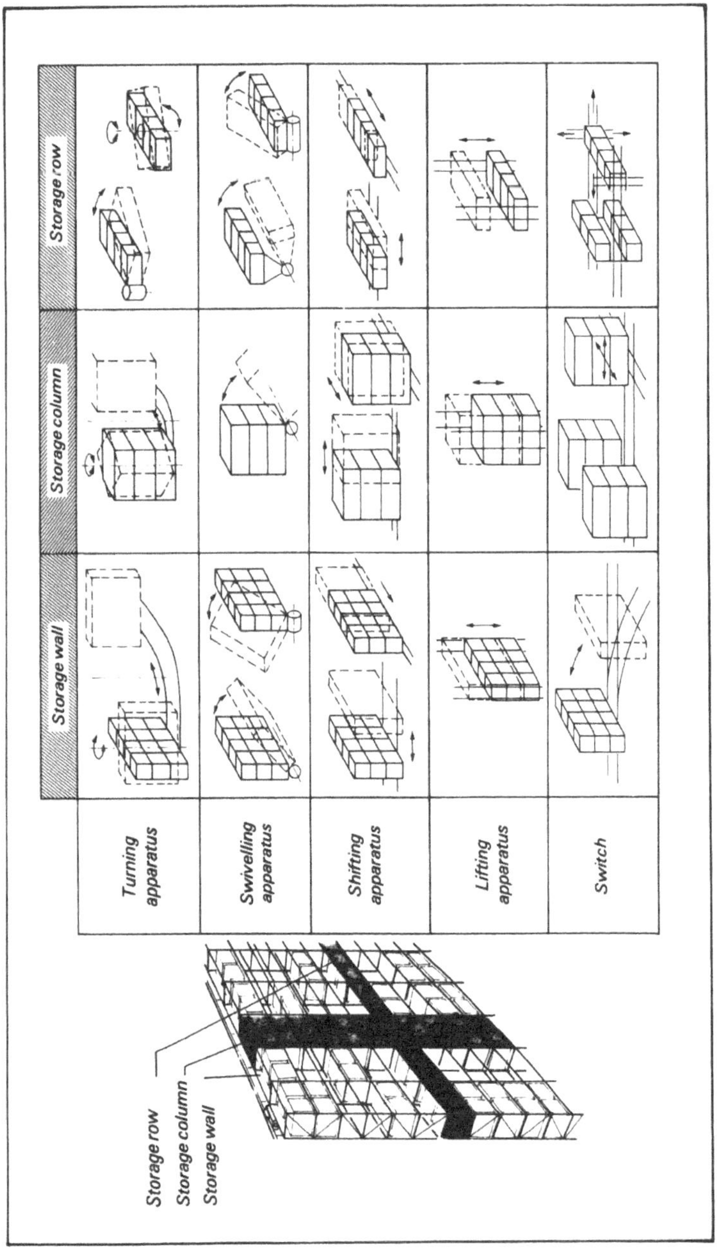

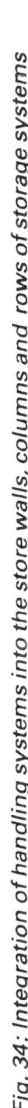

Fig. 34: Integration of handling systems into the store walls, columns and rows of storage systems

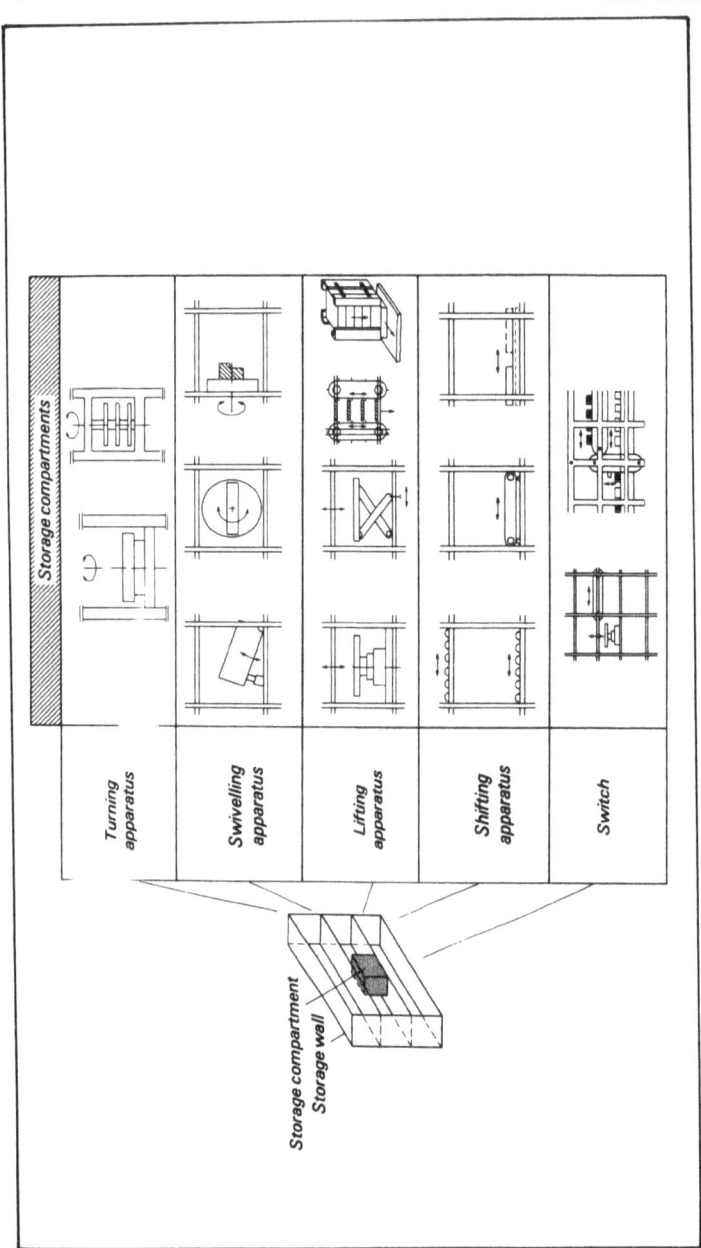

Fig. 35: Integration of handling operations in the store compartment

nents can be present in a different order. First of all, the components may be taken up unordered; that is, place, orientation and quantity are not strictly defined. With increasing precision of these component characteristics the level of ordering increases. Thus the orientation of components may be optional, in slightly preferred orientations, or fully ordered in specific orientations.

As far as quantity is concerned, there are components which are present in quantities though not strictly defined; and components which are collected in groups or separated from each other.

According to the position of a component which is picked up with other components, the above alternatives exist in relation to the type of power input. At the same time for ordered components there is the possibility of taking up varied positions within a stack or pattern.

The configuration of component pick-ups for several components (such as dual grippers) allows the (short-term) storage of components without any special storage facility.

In the integration of the handling sub-operation 'movement', the operating units of a storage system comprise:

• Partitioning wall.
• Store compartments.
• 'Readiness' stations.
• Support apparatus.

They must be specially considered as different possibilities rest for the division of handling sub-operations on these operating units (Fig. 33). Each of the operating units of a storage system can undertake handling sub-operations. Thus various operating procedures arise which can carry out the handling sub-operations through different operating units.

These possible operating procedures are presented in the vertical margin of Fig. 33. Since it follows that in the performance of a handling process either pick-up or placing of components, or movement of component carriers is taking place, the representation of these sub-operations has been omitted from Fig. 33.

In Fig. 33 the alternative methods of performing handling sub-operations (listed at the top of the diagram) are set against operational procedures. Depending on the requirements of the handling operations, turning, lifting, swivelling or storage apparatus and so on, can be used.

By the use of these items of equipment, the various handling operations can be integrated into the storage system. It is seen that in most cases only a part of the handling operations can be carried out by the integration of one of these items of equipment into the conveying system. In order to carry out on demand all the handling operations, the necessary combinations of the above pieces of equipment must be integrated into the conveying system.

According to the requirements of the handling operations, turning, swivelling, shifting and lifting movements, as well as switches and storage positions, can be integrated into the operating units (Fig. 34).

When moving the larger elements of a store, such as the partitioning walls, column or aisle, it must be considered that simultaneous movements of several components will always be involved; that is, it is not possible to handle individual components in this way. Thus the operation of changing quantities can only be performed in a limited way, in that the respective component quantity of a partitioning wall, column or aisle is collected together or branched off. In contrast to this, the other handling operations can be performed on single components, in that the store element moves as a whole in such a way that the relevant single component behaves in the same way. One example is the turning of a store row through 180°, where the orientation around the turning axis and the position in relation to the turning radius is changed. The sequence of components within the store row is thus reversed.

The movement of larger elements of a store provides a favourable time solution, since the movement process does not have to be repeated for each component. However, this solution comes up against technical difficulties in view of the large masses to be moved. It can only be used in a limited way in single unit production and medium production runs because of the undifferentiated handling of single components.

Its use, however, is possible if appropriately small stores are constructed. One example of this is the revolving store, e.g. paternoster or moving shelving, in which, through this movement of complete store rows or partitioning walls, components can be positioned and separated.

As to movement within store compartments, that is, relative to a single component or component carrier, there are alternatives as shown in Fig. 35. The integration of handling operations in store compartments seems only to be sensible in very short-term store processes, otherwise an efficient time use of the apparatus cannot be made.

In larger stores where components are moved less frequently as single items it is an advantage to centralise the handling sub-operation. Here, store support equipment or the waiting stations can be used (Figs. 36 and 37).

As with store compartment, the handling operations can be integrated through turning, swivelling, lifting and shifting apparatus into the support apparatus and the idle stations. The inclusion of store positions is only sensible in support equipment because of the existing direct gripping facilities on the store compartment, if, because of longer driving times, the movements of the support apparatus must be optimised according to sequence.

The construction of store positions at waiting stations is necessary if the internal support apparatus and the external conveying and manufac-

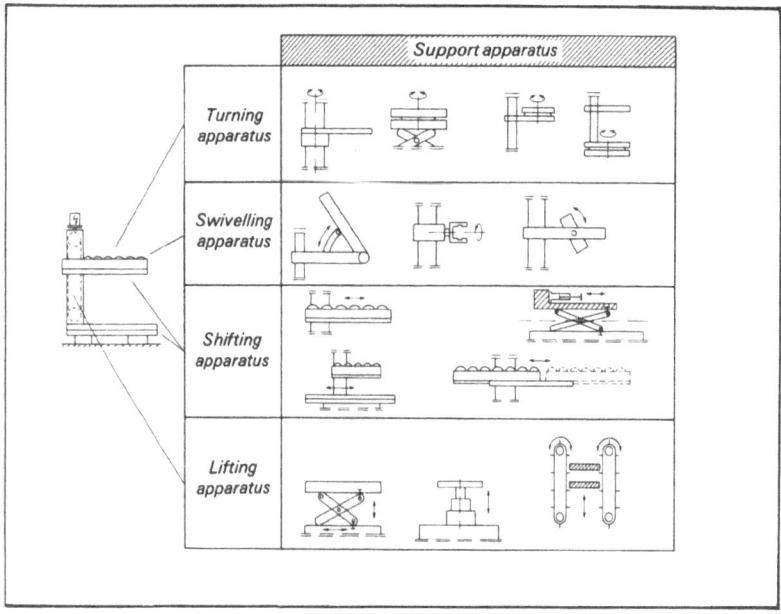

Fig. 36: Integration of handling operations in support apparatus

turing systems show differing time patterns.

For the integration of the handling sub-operation 'movement' in support apparatus and waiting stations, there is also the possibility of performing the movement processes with the whole apparatus instead of only the component pick-up. This is, because of the combined mass and dimensions, only useful for smaller materials handling systems and then will be achieved mainly by simple turning and shifting.

Two examples of an integration of handling operations in storage facilities are shown in Fig. 38 in terms of waiting stations with small store contents.

In this diagram short term stores are represented in which, through simultaneous swivelling of all components, all handling operations can be performed. To this end, specific pick-ups (clamping, positioning aids) for components or component carriers are necessary.

With both these stores (horizontal and sloping turning axes) the orientation of components, among other things, can be altered. So can, for example, the components in horizontal stores. After the store has been turned, the components are in a vertical orientation.

Possibilities for integrating the sub-operation of 'storage' in the

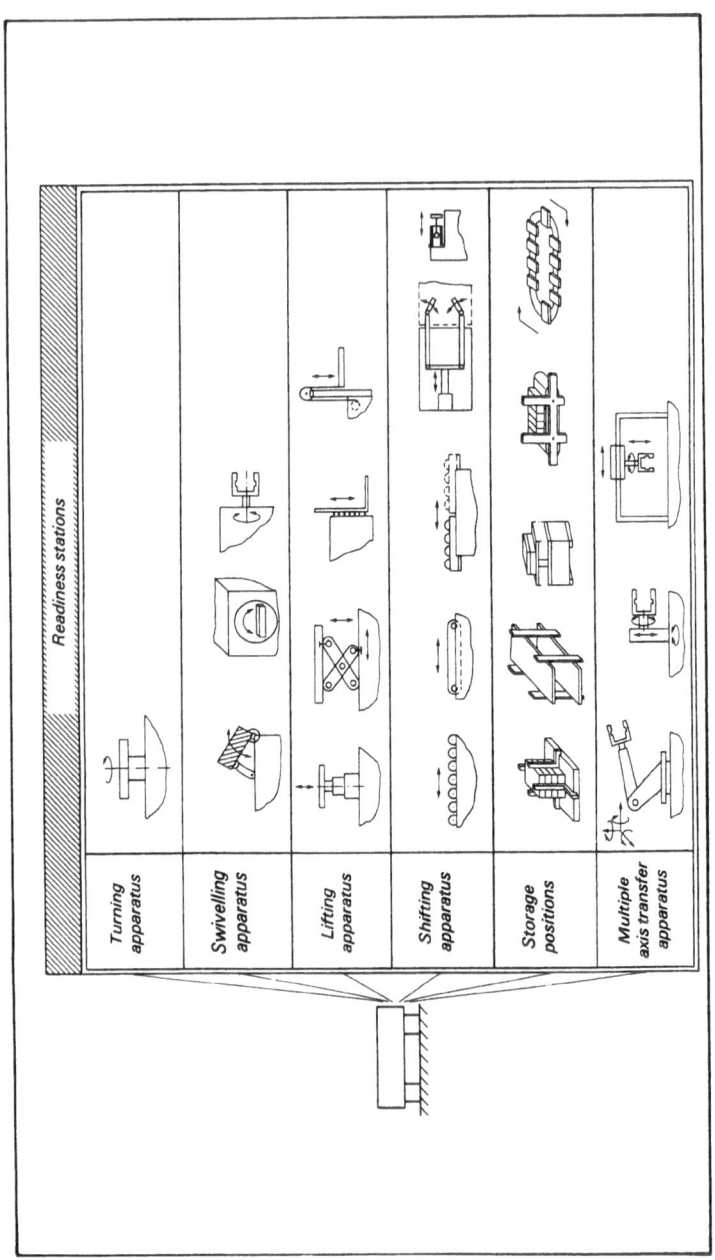

Fig. 37: Integration of handling operations in the waiting stations

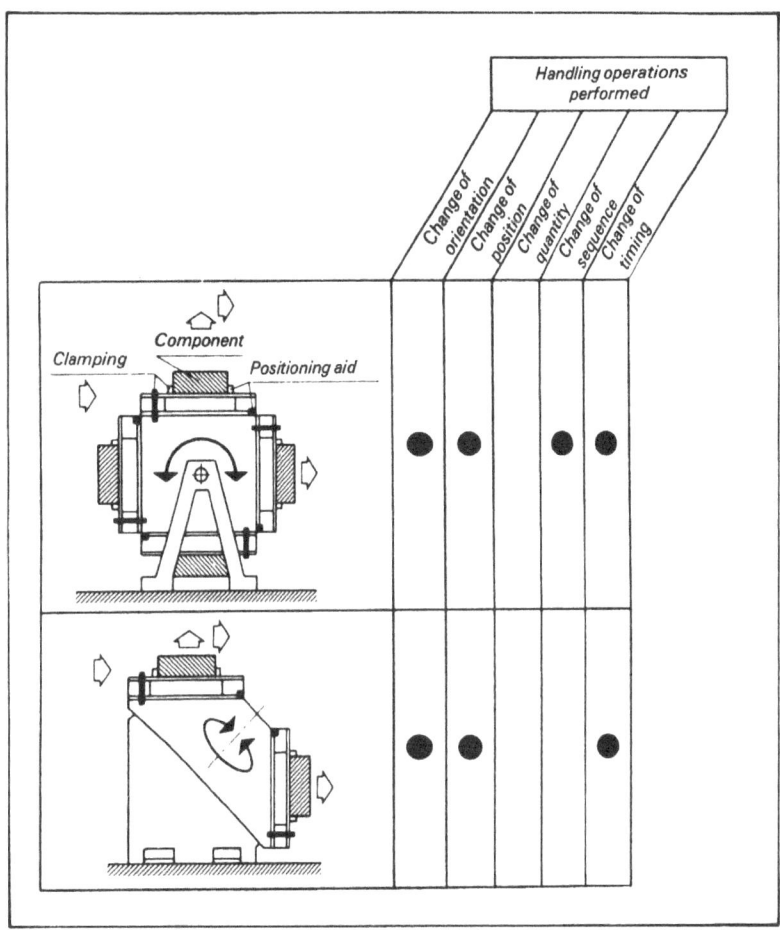

Fig. 38: Possible waiting stations (short term stores) with integrated handling operations

waiting station of a store are shown in Fig. 39. All forms of store are particularly suited to this operation in appropriately reduced size.

Through the inclusion of transfer equipment (e.g. rollers) in the top example, the system in this store can achieve various positions. The same thing applies in the bottom example, though for space-saving reasons the store position here would be arranged one above the other, thus necessitating lifting apparatus.

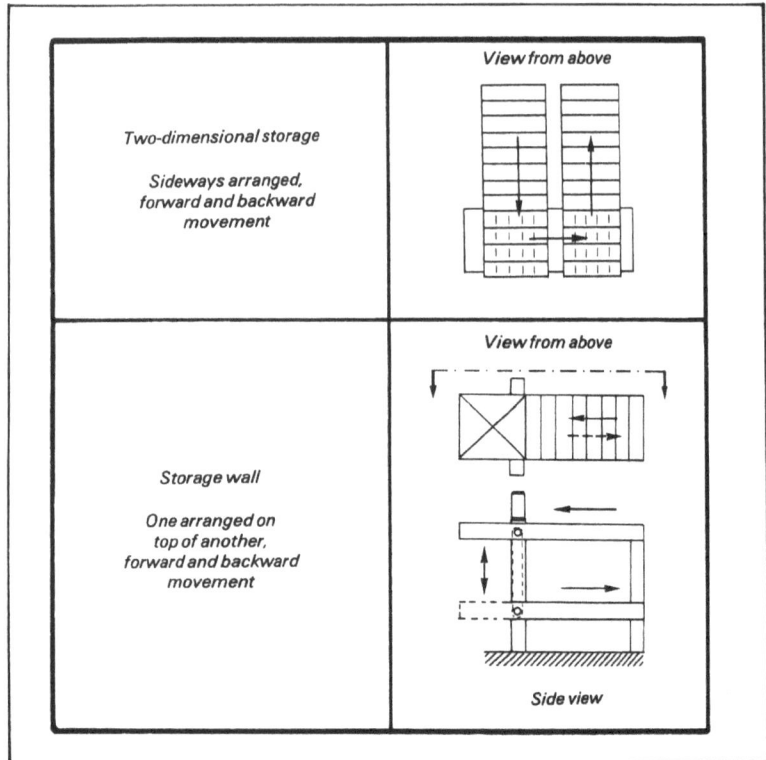

Fig. 39: Waiting stations with integrated handling sub-operation 'storage'

Integration of handling operations in conveying facilities. When integrating the pick-and-place handling function, similar alternatives exist in conveying facilities as those which applied in storage facilities.

Proceeding in the same way as for storage facilities, a determination of the alternatives for integrating the moving and storage handling sub-operations requires operational procedures to be established. These reproduce the different alternatives for a division of sub-operations among the operating units:

- Conveying track.
- Component pick-up.
- Transfer station.

Alternatives for integrating movement and storage mechanisms can be compared with these operational procedures (Fig. 40).

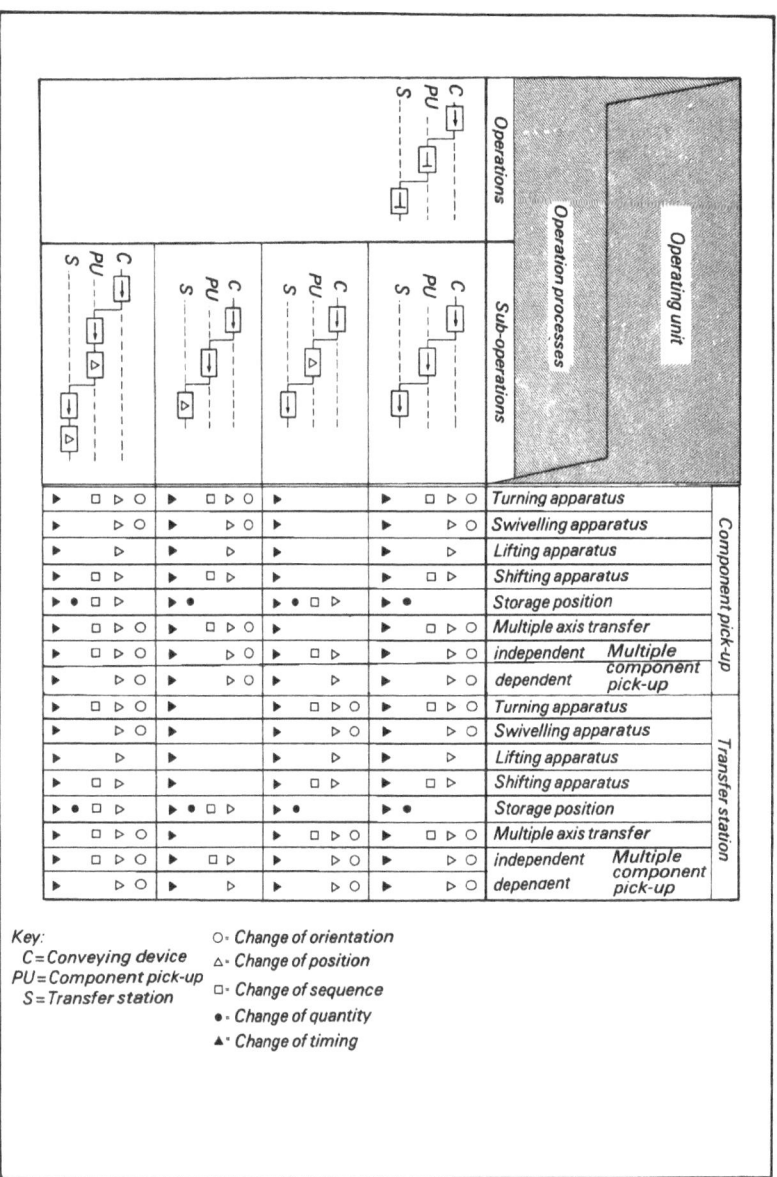

Fig. 40: Alternatives for the integration of the handling sub-operations 'moving' and 'storage' in conveying facilities

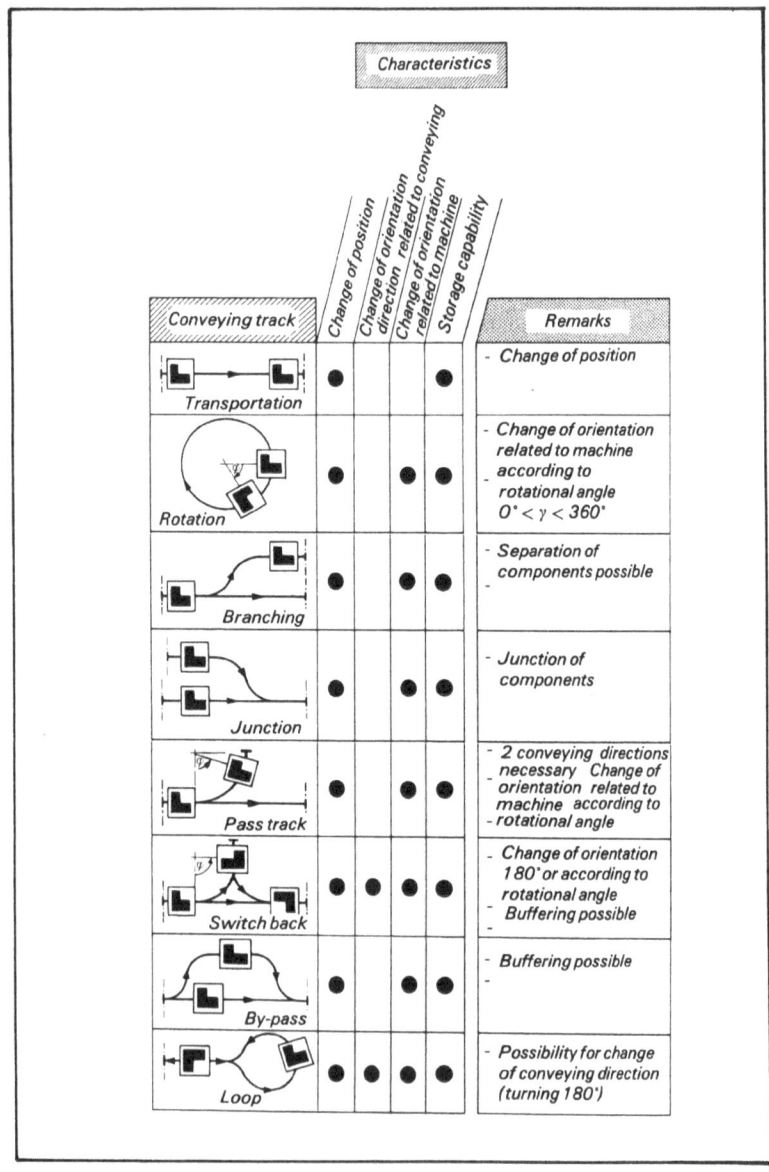

Fig. 41: Influence of the conveying track on handling

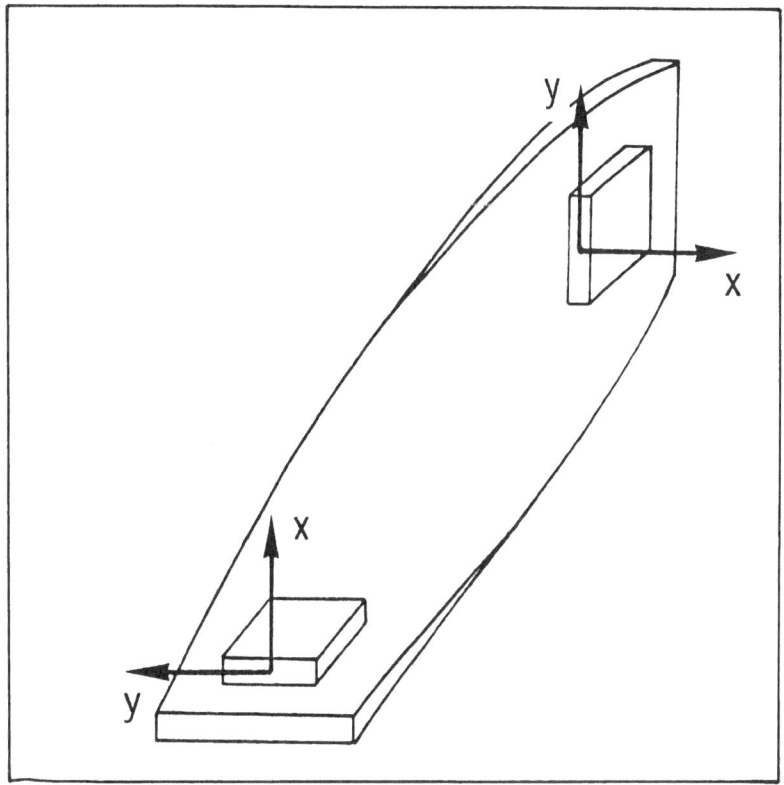

Fig. 42: Orientation change is achieved by rotation in the direction of movement

As with storage facilities, practicable handling operations can be co-ordinated with the above alternatives.

The particular nature of conveying systems makes it possible to integrate handling operations in the conveying track because no further movement or storing mechanism is necessary. The storage capability of a conveying track is always given and depends on the number of devices used or the pick-up capability for components and component carriers (rolling track, conveyor belt)(Fig. 41).

The arrangement of a conveying track involves not only simple transposition and rotation, but also the possibility of branching and other junction points. Here, further track elements can be developed, such as passing sections, right angle turns, bypass sections and loops.

With passing sections, right angle turns and bypass sections,

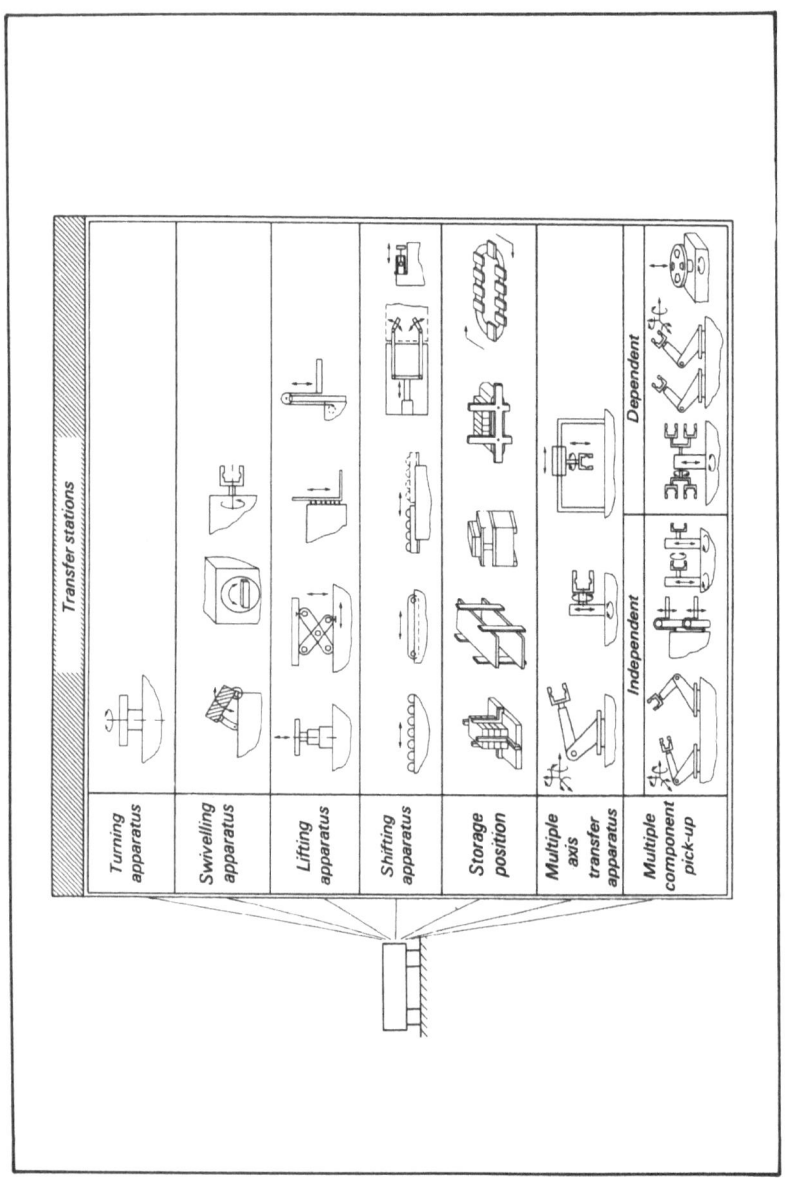

Fig. 43: Integration of handling sub-operations 'pick-up' and 'storage' in the stations of conveying systems

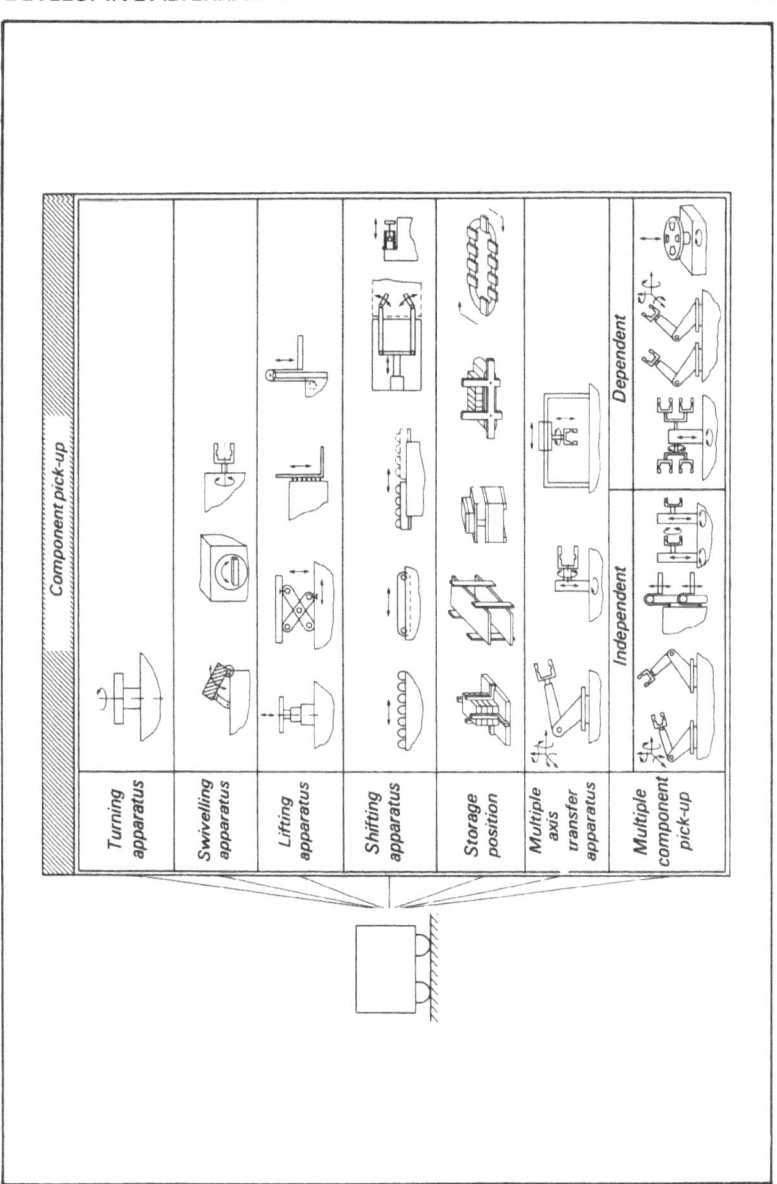

Fig. 44: Integration of handling sub-operations 'pick-up' and 'storage' in the conveying devices of conveying systems

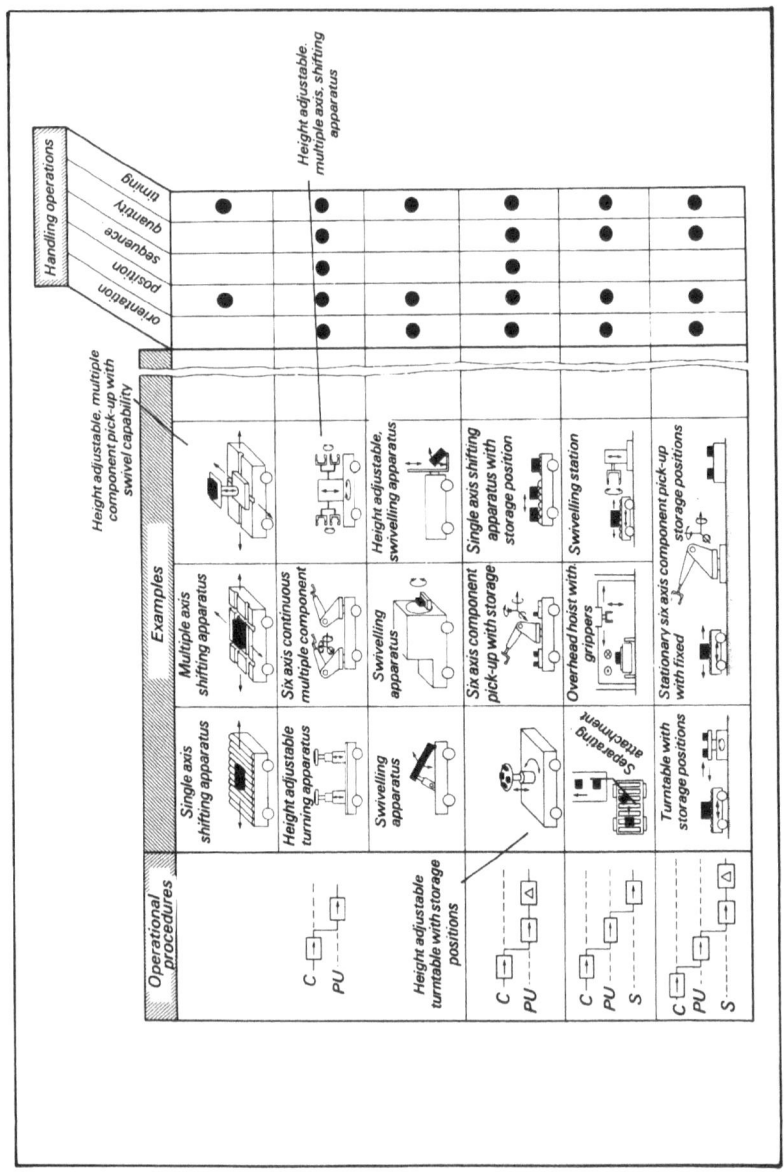

Fig. 45: Examples of alternative methods of integrating handling
operations in conveying facilities

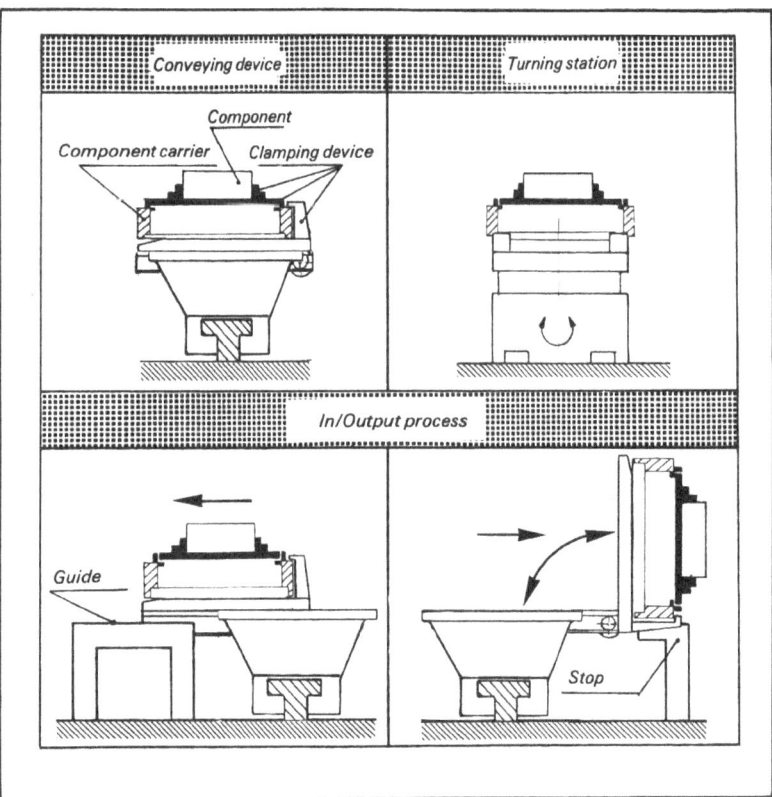

Fig. 46: Monorail with integrated handling operations

components can be stored so that they do not affect other material flowing on the track. The sequence of components can thus be varied by this arrangement of track element.

Depending on the angle of turn required, an orientation change is possible for the component with every curve (rotation) on the conveying track. The orientation of the component is reversed in terms of the conveying direction by right angle turns and loops.

Fig. 42 shows how an orientation change can be achieved by rotation in the direction of movement. This type of equipment for a change of orientation is well-known, for example, vibrating spiral conveyors for small components.

Again, turning, swivelling, shifting and lifting apparatus, as well as

Handling sub-operations / Operating units	Picking up/placing ⌐ ¬	Moving →	Storing △
Measures to integrate handling operations	Integration of additional component pick-ups	Use and expansion of movement possibilities	Integration of additional storage positions
Spindle	Milling Machines ● Exchange of component grips ● 2. Component pick-up	● Turning, shifting in manufacturing or transfer position	● Lathe
Component pick-up	Lathe ● Picking-up with headstock and steady rests	● Movement in the component pick-up (e.g. swivelling feed)	● Several component pick-ups
Component changing system (pallets)	● Multiple clamping	● Integration of additional turning and swivelling units	● Pool systems
Mounting tables (milling machines)	● Several component pick-ups on one mounting table	● Turning and swivelling apparatus ● Integration of additionalsystems	● Several mounting tables with several component pick-ups
Werkzeug-wechselsystem	● Gripper in storage position ● Additional gripper	● Additional movement axes in the tool changer ● Movement of component with storage cycle ● Movemement	● Components in storage position

Fig. 47: Alternative methods of integrating handling sub-operations in turning and milling machines

additional storage positions, can be used to integrate the movement and storage handling sub-operation into the stations and conveying devices of a conveying system (Figs. 43 and 44).

The technical alternatives for integrating the above-mentioned equipment are identical for conveying devices and stations. By combining several movement axes, multiple axis transfer apparatus can be developed. These can handle several components or component carriers simultaneously when used in conjunction with component pick-up. The provision of multiple component pick-ups allows storage of components without requiring dedicated storage positions. In this way, the required movement and storage process can be carried out simultaneously or independently from one another.

Fig. 45 shows examples of integrating handling operations in conveying facilities using individual floor-standing vehicles.

One alternative method of integrating handling operations in conveying system transfer stations is to transfer components by separators or stations under which conveying devices can pass, since, in this case, no further movement mechanism is needed. In order to place the component in a defined orientation, defined pick-up geometry is needed in the object to be conveyed. A typical example is the standardised component carrier.

The integration of multiple axis transfer apparatus usually requires linking an industrial robot with a conveying device using the appropriate automatic equipment control. Fundamental research has already been done in this field[72].

A further example of a conveyor system using integrated handling operations is represented in Fig. 46 with a floor-standing monorail.

Objects to be conveyed are either single components or component carriers transporting several components. With the aid of the conveying track, and additional shifting and swivelling apparatus, the conveying system can change the position and orientation of the object to be conveyed. Thus, components can be exchanged between conveying facilities and storage or manufacturing facilities. The swivelling of components, as well as the distribution of component carriers, is limited by direction in this system at any given time (only possible at the end of one side). The use of a specially adapted track section, such as a loop, can compensate for this disadvantage however. By using additional turning stations, onto which the component carriers are transferred, components can be brought into positions at right angles to each other with the help of the conveying system.

To carry out short sideways movements, such as required when placing a component carrier in position or inserting a component in a machine tool clamping equipment, a two-way load pick-up can be useful. To achieve very accurate positions, this load pick-up is equipped with guides and blocks.

Component quantity changes, through collection and separation processes, can be achieved by using several of the above mentioned conveying devices. The handling operation 'sequence and time order change' is possible through the use of a specially adapted track section (e.g. siding tracks, bypass tracks).

Integration of handling operations in manufacturing facilities. In the integration of handling operations in manufacturing facilities, oper-

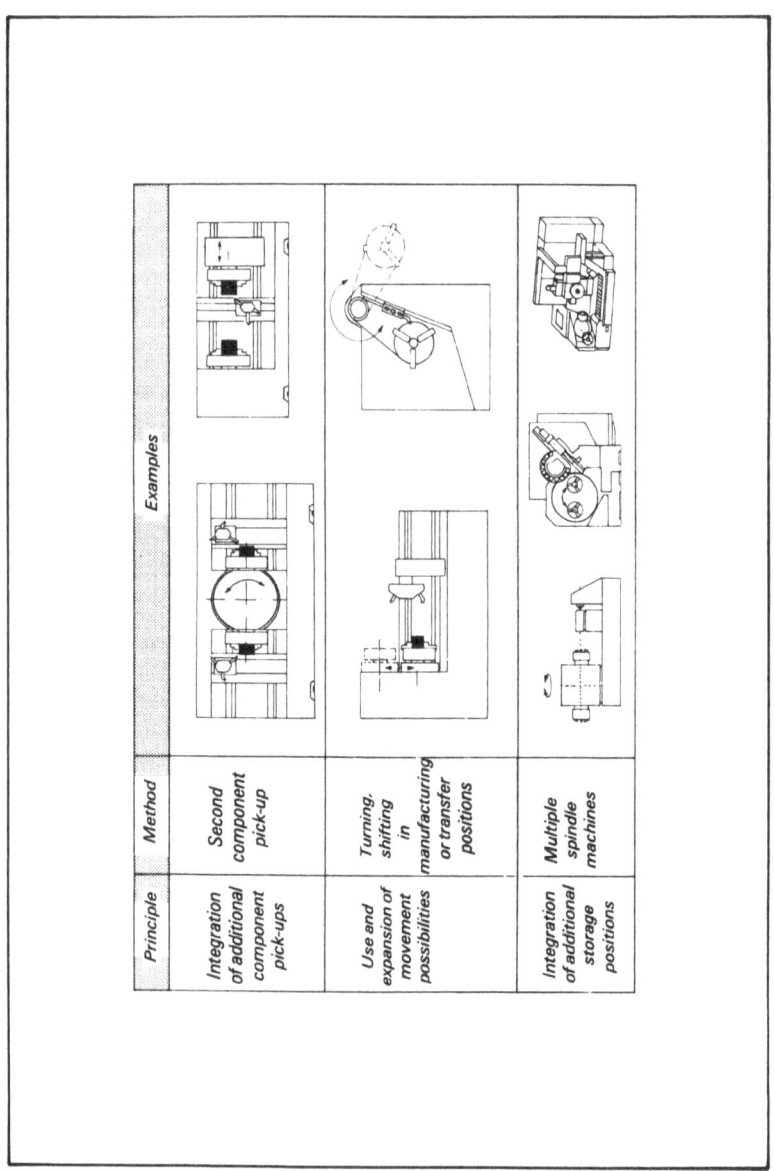

Principle	Method	Examples
Integration of additional component pick-ups	Second component pick-up	
Use and expansion of movement possibilities	Turning, shifting in manufacturing or transfer positions	
Integration of additional storage positions	Multiple spindle machines	

Fig. 48: Alternative methods of integrating handling sub-operations in the shafts of turning machines

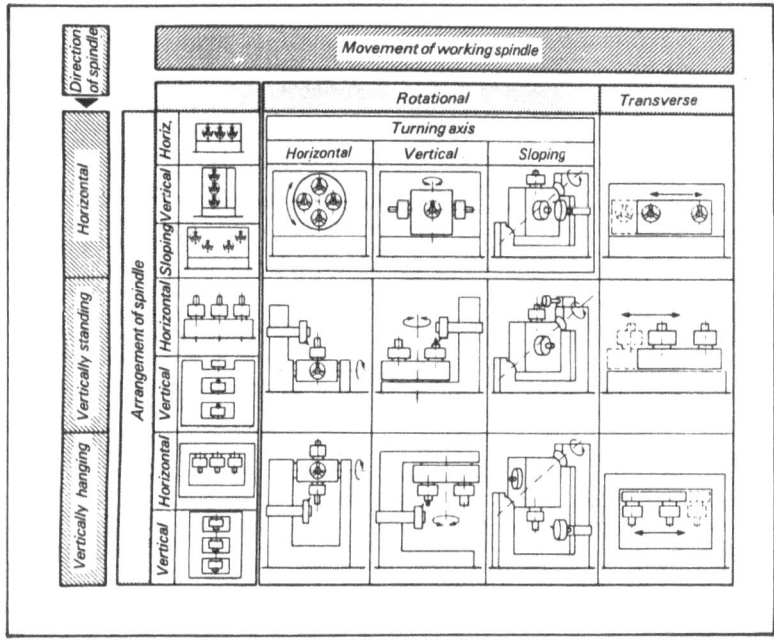

Fig. 49: Alternative methods of integrating the handling sub-operations 'storage' and 'movement' in turning machines using multi-shaft machine concepts.

ational procedures can once again be established in which operating units are arranged to carry out individual handling sub-operations. It is shown that the measures to integrate handling sub-operations can be summarised as measures:

- To integrate additional component pick-ups.
- To use and expand the alternative methods of movement of existing operating units.
- To integrate additional storage positions (Fig. 47).

In the integration of handling operations, it is useful to bear in mind the fact that many operating units in a manufacturing facility already carry out single handling sub-operations. In this way, e.g. in the case of tool change systems, all handling sub-operations are already being performed – admittedly for a selected range of objects to be handled.

For the systematic determination of technical solutions for integrating handling operations in manufacturing facilities, once again the individual operating units – shaft, tool change system, component pick-up,

mounting table and pallet change system – can be investigated.
The alternative methods of integrating handling sub-routines into the
shaft of turning machines is shown in Fig. 48 [73].

The 'pick-and-place' handling sub-operations are undertaken in a
turning machine using component pick-up on the shaft. The integration
of additional component pick-ups means, therefore, either putting a
further component pick-up on the shaft or setting up several shafts with
individual component pick-ups.

To integrate the handling sub-operation 'movement', the shafts can
be turned vertically on the shaft axis or shifted. To increase storage
capability (handling sub-operation 'storage') similar multi-shaft
machine concepts can be used. There are many alternatives here (Fig.
49).

The various machine concepts shown are differentiated mainly with
respect to the direction of the shaft and its movement possibilities.

As well as increased store capability, machine concepts involving
vertical shaft direction can use gravity in the presentation and delivery
of components. Furthermore, in this case swarf falls automatically out
of the workspace, thus minimising the danger of encroachment on the
pick-up of a component (see also Chapter 6).

With rotary or transverse shafts the movements can reach beyond the
working area of the machine, so that accessibility to the shaft is also
increased.

The machine tool change system can also be used as a further
operating unit to carry out handling operations in turning machines
(Fig. 50). With tool magazines the handling sub-operation 'storage' can
be technically divided from the equipment for 'pick-up, placing and
moving', in that separate component pick-ups and movement mechanisms are provided for this. This is recommended if the characteristics of
the components relevant to handling (dimensions, weight, shape etc.)
diverge strongly from the appropriate characteristics of the tool or
require the minimisation of idle time by simultaneous change of tools
and components.

Like the tool change system, the existing component pick-ups,
headstocks and steady rests in the turning machine can be used to carry
out handling operations. It is assumed here that, first, the headstock
and/or steady rest is expanded using component pick-up, (e.g. triple jaw
chuck,self-centering clamping rollers) and, second, that the movement
possibilities of these devices are increased (Fig. 51).

The handling of components with the aid of headstocks is adapted –
through the application of power to one side of the turning piece –
especially for short components. In the manufacture of longer items
handling cannot be carried out simply by the use of headstocks. Here
the steady rest 'which will in any case often be necessary for slim

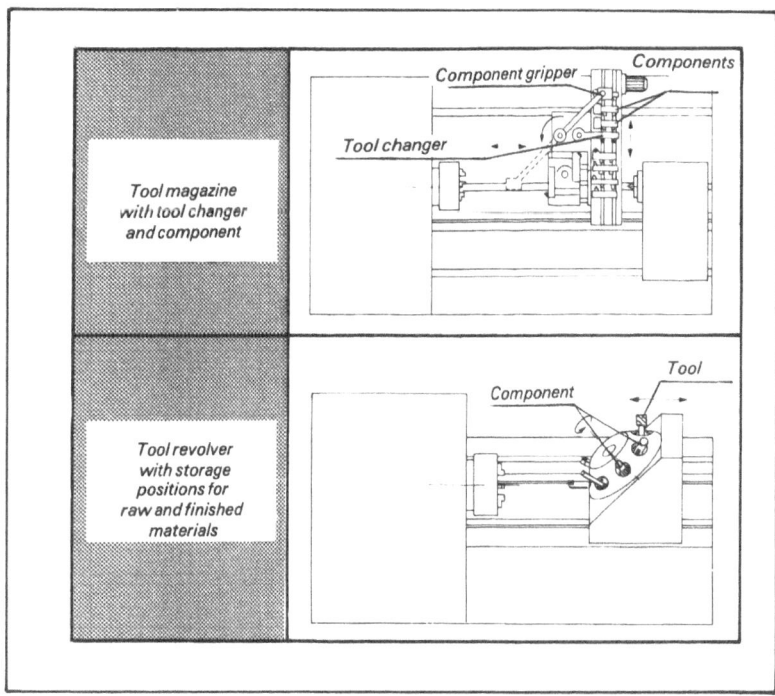

Fig. 50: Alternatives for integrating handling operations in manufacturing facilities through the use of shifting and storage apparatus for tools, with specific reference to horizontal single shaft turning machines

components' (such as axles) can be used for position change.

Through the combination of headstock and steady rest for component handling, a manufacturing facility is able to store components to a small extent. It is assumed here that the headstock or steady rest is not required for the support of the manufacturing operations during the storage time of the components. Thus the point in time when components are presented or delivered is freely available for the duration of component machining. In addition, if either headstock or steady rest is equipped to carry out turning movements (180°), re-chucking processes for machining a second side of the component can be completed for components whose turning circle does not exceed the dimensions of the machine's workspace.

The particular advantage of using headstock or steady rest to carry out handling operations lies in the advanced centering of the component on the turning axis. In this way components can be inserted in the

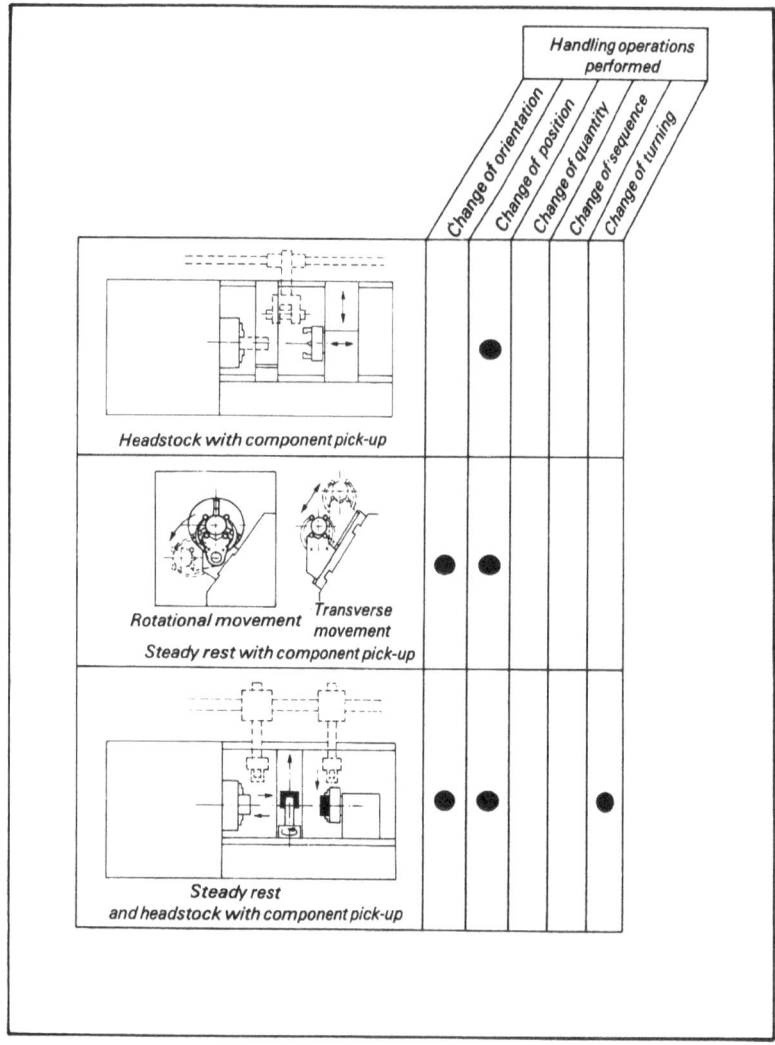

Fig. 51: Use of headstock and steady rest to carry out handling operations
on single-shaft turning machines

component pick-up without wasted movement.

The alternative methods of integrating handling operations in milling
machines are shown in Fig. 52.

Expansion of the capacity of the shaft to carry out handling

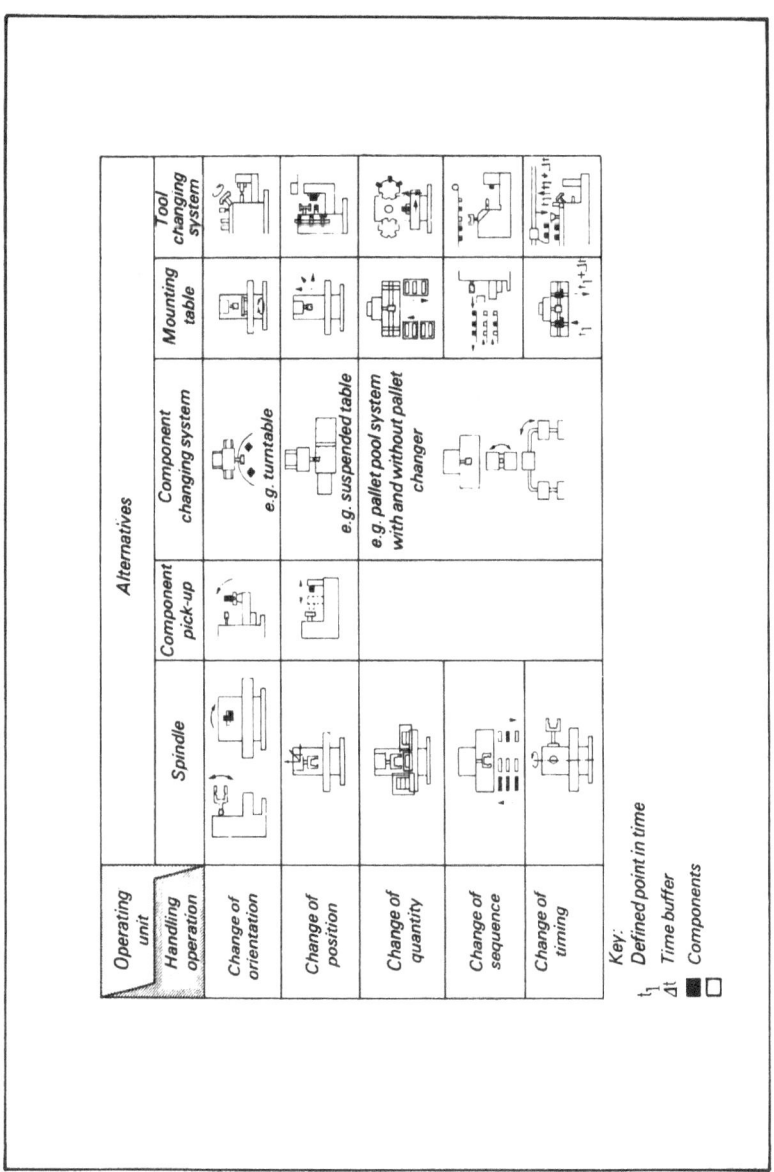

Fig. 52: *Alternative methods of integrating handling operations in milling machines*

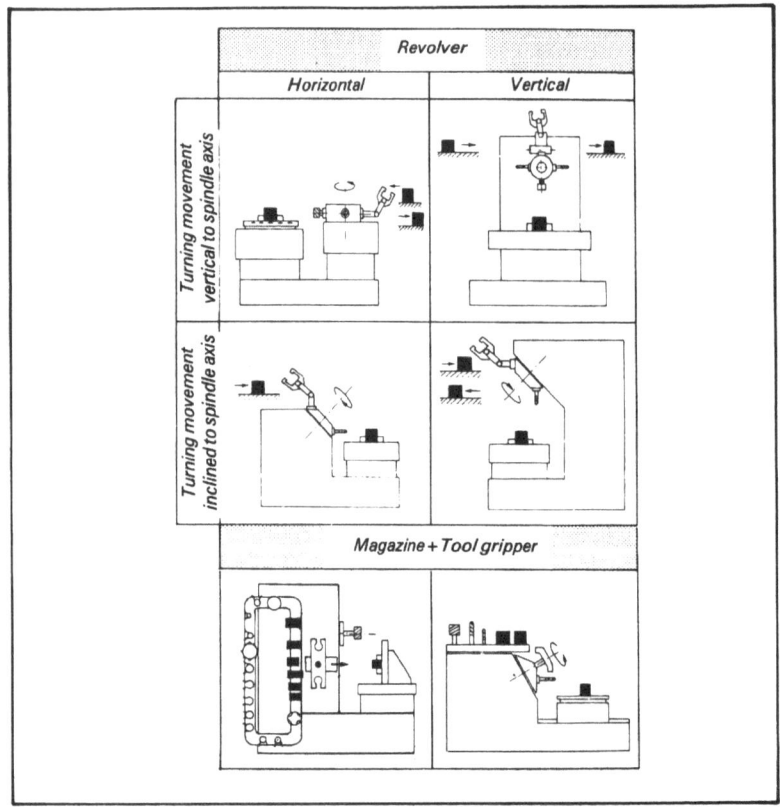

Fig. 53: Component handling on milling machines with the aid of
toolchange apparatus

operations requires first of all the integration of a component pick-up in
the shaft. To do this a gripping device similar to a tool can be used. The
gripping device can be stored in the tool magazine, assuming that this
magazine is large enough. Also, free-standing gripping magazines are
possible. To change a gripping device on the shaft, the transfer of
energy to drive the gripping devices and the transfer of information to
control the gripping processes – for example, with the aid of plug-in
connections – must be made possible.

Another alternative method of integrating a gripping device in the
shaft of a milling machine is the introduction of gripping devices which
are fixed with the shaft and stay there after carrying out handling
processes. One possibility is, for example, the power-driven grippers as

found on bar feed turning machines. As any change of gripping device involves, in this case a machine change-over process, this alternative is suitable mainly for component ranges with the same or similar gripping conditions. This is particularly so if, for example, rotation-symmetric parts or components with specific, constant component characteristics for gripping in the form of cams, ridged grips or drill holes, are being handled. A special problem with this kind of 'built-in' gripping device is the danger of soiling the gripping device with machining waste (chips, clippings) and cooling/lubricating agent since the gripping device is always in close proximity to the working area of the tool. Furthermore, there is the possibility that parts of the gripping device could collide with the component or apparatus.

The high precision and stiffness of the machine tool involving fixed gripping devices is an advantage in handling processes requiring the exact positioning of a component. In general the operating speed of the machine tool in rapid motion must be fundamentally increased, so that with this alternative for carrying out handling operations, cycle times comparable to those achieved by free-standing handling apparatus can be attained. At present, the achievable high speed of milling machines lies at about 5 and 10m/min (max. 15m/min), while flexible handling equipment speeds are about a factor of six times higher, and can achieve 0.5–1m/s.

To carry out the handling sub-operation 'movement', a combination of shaft rotation and advance can be used. Thus with the aid of shaft rotation the orientation of a component can be changed, for example, as in re-chucking. For this the shaft must be equipped with specially adapted turners and clamps, to allow the precise position and orientation of the components to be determined.

Milling machines with swivel shafts can also change the orientation of a component through the swivel movement.

To perform the handling operations 'quantity and sequence change' it is necessary to have storage potential within the movement zone of the shaft or multiple gripper. The integration of the handling operation 'time order change' into the shaft of milling machines can be achieved in a similar way to the multiple shaft turning machines, by the addition of further shafts.

The integration of handling operations in the component pick-ups of milling machines involves the inclusion of movement and storage equipment in the clamping device. Even then this is only useful if additional movement axes are integrated, exceeding existing movement possibilities for this machine. This means, however, that the integration of handling operations in these installations only happens in special cases.

In cases where the component range has a higher requirement for

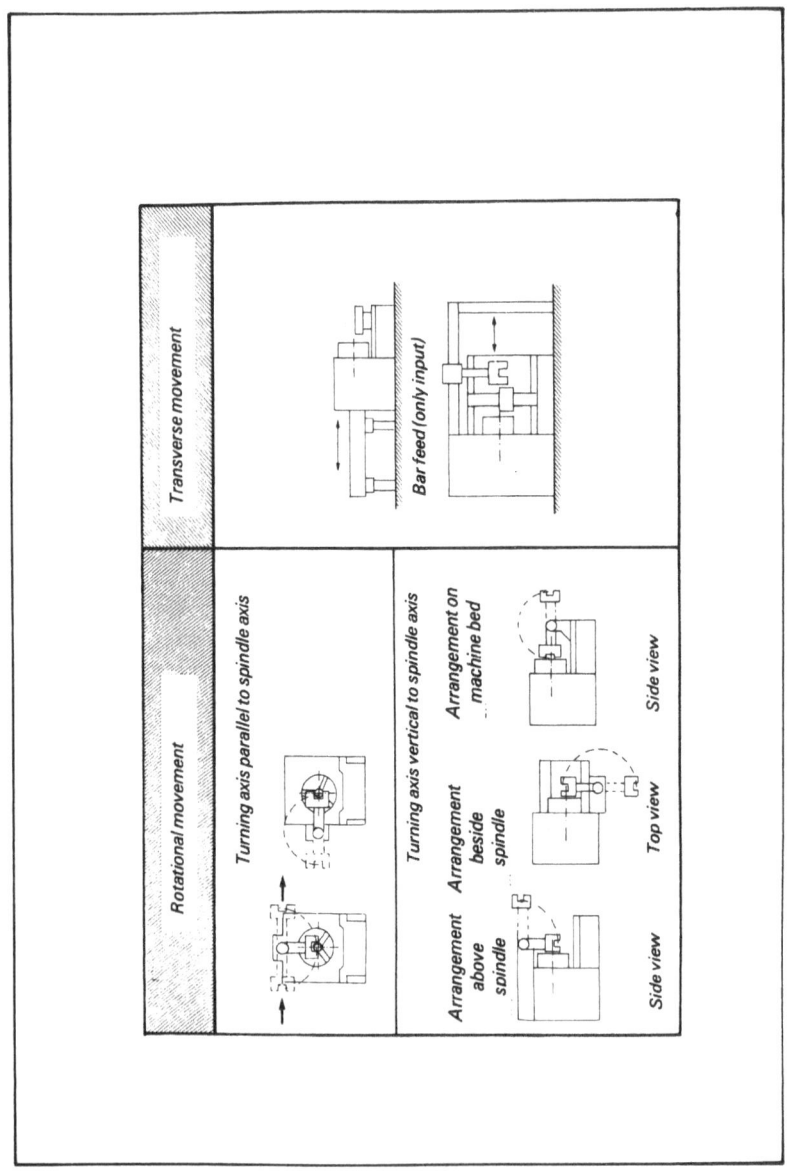

Fig. 54: Integration of additional handling apparatus to perform the
handling sub-operation 'movement' in the example of a turning machine

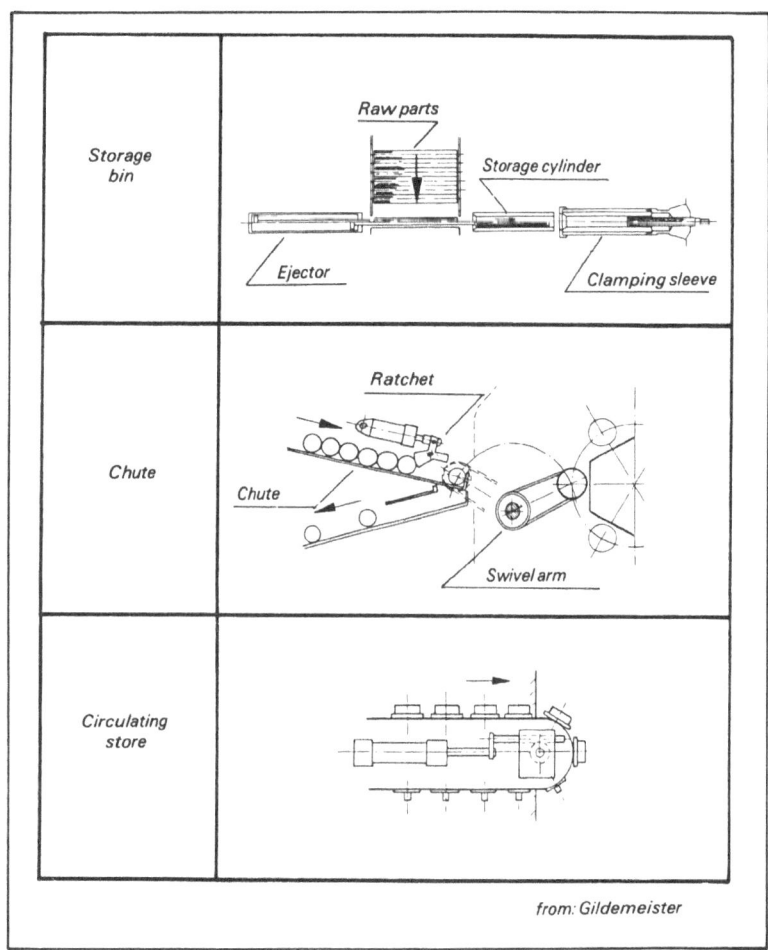

Storage bin

Raw parts

Storage cylinder

Ejector

Clamping sleeve

Chute

Ratchet

Chute

Swivel arm

Circulating store

from: Gildemeister

Fig. 55: Types of component storage for machine tools

supplementary movements, the machine tool will be fitted with àdditional movement axes for turning, swivelling, lifting and shifting. This fitment takes place before the existing clamping device is provided with the sometimes necessary movement mechanisms and the related drives, measurement systems and controls.

The alternative methods of integrating handling operations with the aid of tool change systems is shown in Fig. 53.

With the performance of handling operations by tool grippers, it is

Chapter Six
Developing production concepts

WITH the aid of the alternatives illustrated for integrating handling operations in production facilities, concepts for production systems can be developed. As already shown, integrating handling operations in storage, conveying and manufacturing facilities follows from the planning of type, number, arrangement and level of automation, etc. of production facilities.

Procedures
Integrating handling operations in production facilities involves an overlap with several storage, conveying and manufacturing facilities. First of all it is helpful to define the areas within the production system to be investigated in order to approach the various distinct considerations which need to be looked at in the integration of handling operation in production plants. The object of this arrangement of production systems in terms of production area is to separate the storage, conveying and manufacturing facilities which, either because of their low levels of automation, or because of excessive component, procedural and production requirements (too great, or excessively long manufacturing times, very short piece-part times etc.), cannot be considered for full automation.

On the other hand, this means that the outlay for planning the whole production system can be divided into a number of comprehensive parts. A fundamental part of the definition of production areas is the clarification of whether manufacturing facilities can be supplied directly by a conveying or storage system or indirectly by machine-related stores (Fig. 56).

By building on the definition of production areas, the handling facilities for pick-up/placing, storage and movement of components can be established. To this end the production facilities must take account of the space conditions and time relationships between each other.

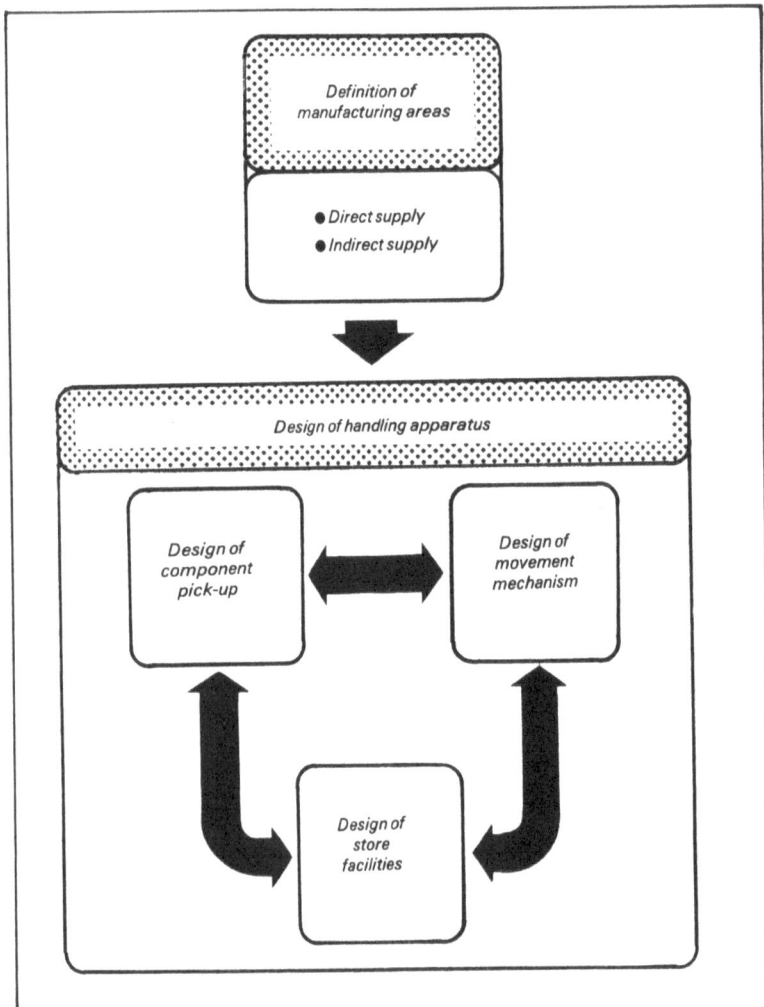

Fig. 56: System for the determination of production concepts taking into account the possibilities for integrating handling operations in production plants

Boundaries of production areas. In investigating possible automated production concepts for a batch of one or serial production, several automation stages can be distinguished which differ in relation to the

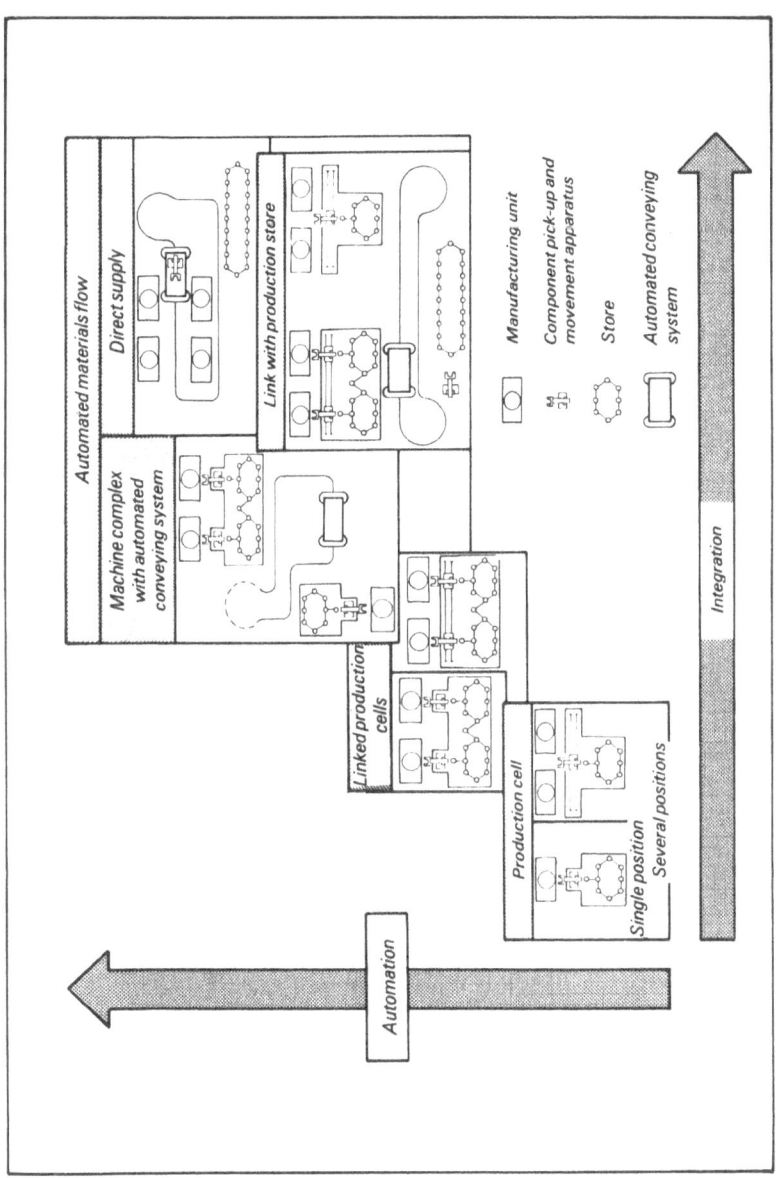

Fig. 57: *Automated production concepts for single unit and volume production*

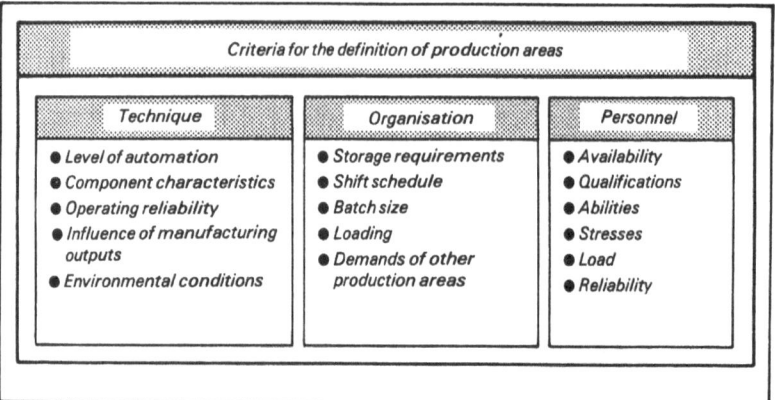

Fig. 58: Criteria for the definition of production areas as a basis for the planning of production facilities with integrated handling operations

level of automation and the extent of connected automated production facilities (Fig. 57).

The simplest form of automation which exceeds the automation of the manufacturing operation, is the production cell. Here the handling operations on a single machine are automated. Machine-related stores, movement mechanisms and component pick-ups are necessary for this.

The next automation stage involves several machines supported by a single handling device. Here the manufacturing facilities are not linked to perform consecutive operations. Since the production cells are not coupled to automated conveying systems, handling conditions between production cells and conveying systems correspond only to the ergonomic arrangement of working conditions for operators.

A linkage of consecutive operations can be achieved using linked production cells. Linkage is achieved in this stage of automation either by a store using integrated movement mechanisms or by the free-standing movement device of a supplementary handling system.

In the next stage of automation the transport of components between production cells and the production store is automated by degrees. The highest level of automation means the end of the production cell, that is, the coordination of machine-related stores in the production store and direct supply from a conveying device. Flexible production systems in the currently known form are good examples of this.

It is important that in all automation stages with indirect supply of the manufacturing facilities, the necessity for machine-related handling systems with component stores and transfer mechanisms persists. With direct supply, machine-related store facilities cease, since the handling

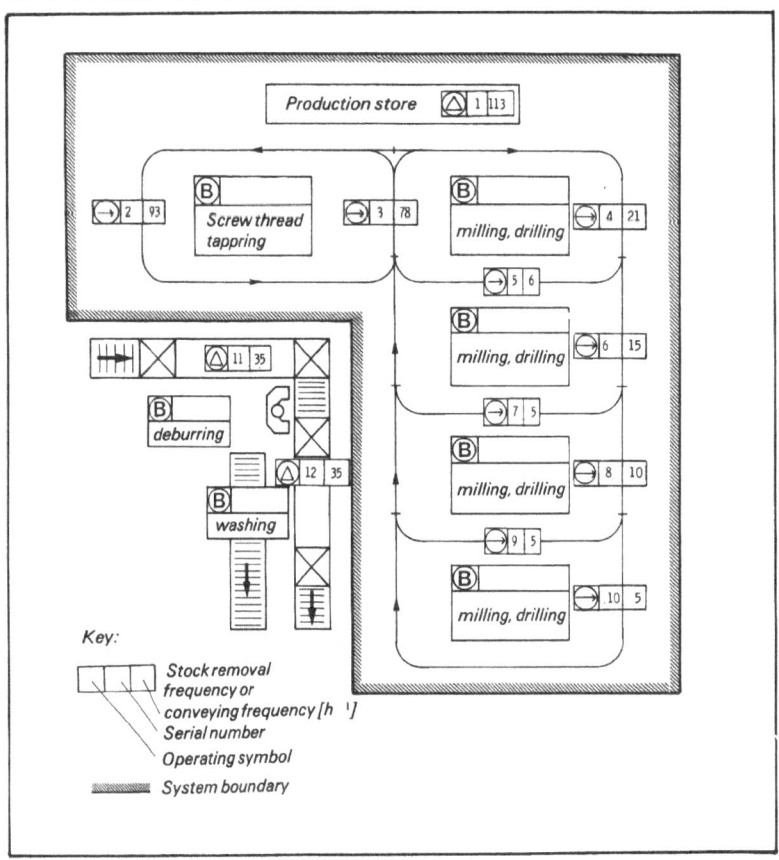

Fig. 59: Example of the rough layout of a production area as a basis for planning a production facility using integrated handling operations

sub-operation 'storage' is taken over by the production store.

The most important criteria in the definition of production areas, which are relevant to integration, are therefore the requirements for storage facilities at the manufacturing facilities and the level of automation of the production facilities under investigation. In addition, technical, organisational and personnel-related aspects are involved in the definition of production areas (Fig. 58).

A division appears in the component range of a production system according to component characteristics relevant to handling since, for example, large variations in size, form or weight require technically different methods of carrying out handling sub-routines.

In addition to the component range, the manufacturing facilities themselves can also force a division of production areas. This happens particularly when the manufacturing process cannot be satisfactorily controlled, so that operators must supervise the manufacturing facilities and they can undertake the required handling tasks.

Further, technological and environmental conditions (such as vibrations, heat and dust) can require certain manufacturing facilities to be separated from a production area using various integrated handling operations.

A further possibility is the definition of production areas according to organisational aspects. For example the production store might operate one shift while the manufacturing facilities run multiple shifts.

Furthermore, it must be considered that lot sizes and delivery times of the components to be manufactured in a production area tend to move in rather similar fashion. Thus it should be possible to avoid the extreme case of having to manufacture and handle components in very low volume and consequently changing handling conditions. However, the decisive factor in these considerations is the loading of manufacturing facilities which must also be ensured after this kind of division of production tasks.

Since production areas with integrated handling operations always occur as closed systems, care must be taken that in the division of these production areas, in certain circumstances, individual production facilities, which cannot be involved in automated handling processes, may to some extent operate together. Then, specially adapted transfer positions to the manually served production areas must be planned, or else the production facilities involved must be removed from the production area using integrated handling operations.

Finally there exists the possibility of defining production areas according to the requirements of personnel. As well as the time availability of personnel for a multiple-shift operation, the qualifications and remaining activities of the personnel are very important. In addition to the stresses and strains arising from personnel activities, the safety of personnel should be ensured in the definition of production areas with automated component handling. This can happen, for example, through a complete re-structuring of workstations with manual support.

Fig. 59 shows an example of the definition of a production area intended for integrated handling operations. The production area includes in the first instance four manufacturing centres to machine diecast metal housing parts. The inclusion of a device to tap screw threads assumes the automation of the screwing and feeding processes. The coupling of the production store implies that this is used only through this considered production area, in order to exclude false positioning and encroachments of other production areas.

The manually operated turning and washing positions were taken out of the production area. In contrast to the manufacturing centres, they present significantly shorter cycle times and thus completely different time characteristics. Because of the short operating times they are also used by other production areas. For safety reasons service personnel are divided from the production area by roller conveyors and stores.

The basis for planning component pick-ups, movement mechanisms and stores in production areas will be shown in the following sections.

Planning component pick-ups. The configuration of component pick-ups is influenced by characteristics of components and production facilities (Fig. 60).

The first task in planning component pick-ups in production plants using integrated handling operations is to determine the number of components to be picked up. This is determined in manufacturing facilities by the number of shafts (turning machines) or the size of the workspace (milling machines). In milling it can happen that components occasionally have several faces to be machined, which must remain accessible. If in this case a multiple re-chucking of the component is not possible for time and efficiency reasons (renewed change-over, renewed mounting), several components should not be allowed on one pick-up otherwise freely accessible machining faces are concealed.

In storage and conveying facilities, determination of the number of components to be taken from a storage compartment or conveying device is differentiated by whether or not a component carrier is involved in storage and conveying facilities. In storage and conveying processes without component carriers there is the possibility of setting up single or multiple grippers. The use of single grippers offers the advantage of being economical. They are always recommended when the proportion of handling time is relatively small compared with storage, conveying and manufacturing time, so that the relative importance of handling processes involving the single gripper is low.

In contrast, with multiple grippers there is the possibility of carrying out handling processes in parallel and the operation is thus faster. In addition multiple grippers can cope with differing geometry, e.g. if the shapes of raw and worked materials differ considerably or if a frequent change of components to be handled is necessary.

When component carriers are used, the number of components to be picked up is governed by the dimensions of the component carrier. Besides, after consideration of interrelated alternatives, additional space for the pick-up elements of the component carrier and the component pick-up of the production facility into which the components will be transferred should be taken into account.

Once the number of components to be picked up is established, the

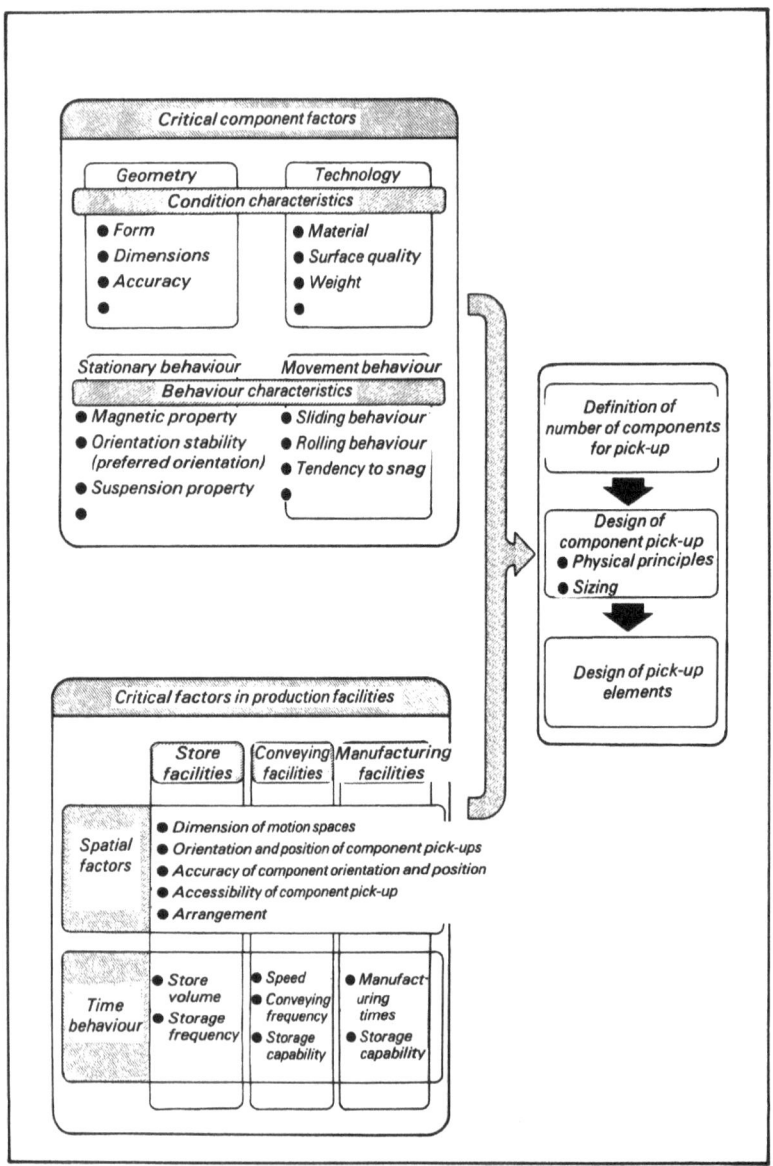

Fig. 60: Criteria and procedures for the technical performance of the handling sub-operation 'pick-up'

component pick-up can be determined according to physical principles actuated by power, shape, material and sizing. For component pick-ups in manufacturing facilities, the need for precision and strength are especially important, if the component remains in the pick-up during machining. Additionally in this case the component pick-up must ensure accessibility of tools to the component.

The configuration of component pick-ups in manufacturing facilities is basically dependent on the clamping position of a component. Since the clamping surfaces of a component are covered by the clamping device these are not accessible for pick-up for transfer to a store or conveying facility (Fig. 61).

The sizing of component grippers for storing and conveying facilities not involving component carriers depends principally on the dimensions and weight of the component. Only in exceptional cases, acceleration in storage and conveying processes necessitates consideration of inertia.

The design of the pick-up elements is determined in the first instance by the component's geometry and the precision of positioning. In manufacturing facilities, the pick-up elements also transmit the required power. Thus accessibility for the handling process should not be restricted.

This means that, in general, clamping elements with a relatively large clamping area are necessary, leaving the way free for the handling process 'movement' through a propulsion of the component after clamping force is removed. Thus the starting and ending of the clamping process must be automatic. Furthermore the component must remain in the component pick-up of the handling apparatus during the clamping process so that it is always under control. Even short slips during the transfer of components means the component is out of control.

To avoid problems which occur in some cases of redundant pick-ups of components transferring between two production facilities, either the component must be very precisely located during the handling process, or the operating unit concerned must be flexible enough so that neither the components nor the production facility are damaged or misformed.

In metal-cutting machine tools, particular attention must be paid to the removal of chips and cooling/lubricating agent from the area of the pick-up surfaces on the component and the pick-up elements of the equipment. In addition, care must be taken to eliminate swarf from the entire component, in order to avoid problems in other production areas.

To increase the accessibility of the component after machining, equipment with special output devices may be used – e.g. mechanical or pneumatic ejectors, or separators. This often results, however, in the components being disordered, so this alternative is only recommended if the following process permits it.

Usually in the configuration of component pick-up in storage and

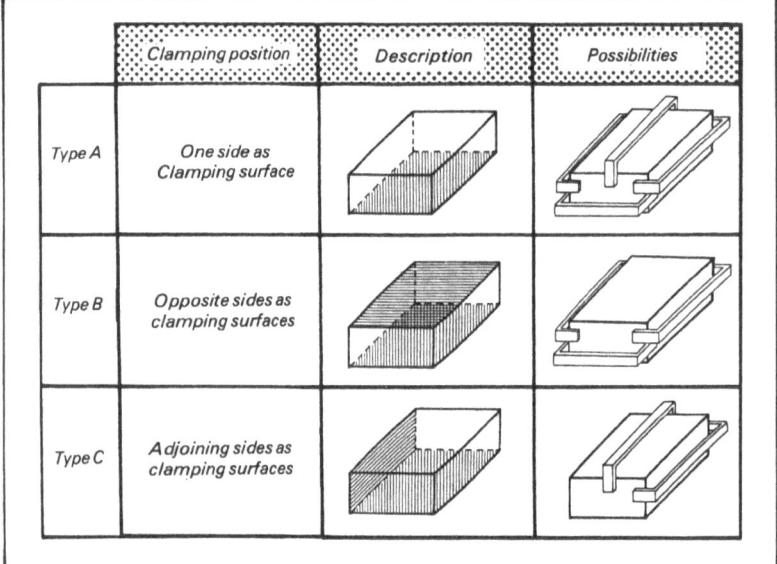

Clamping position	Description	Possibilities
Type A	One side as Clamping surface		
Type B	Opposite sides as clamping surfaces		
Type C	Adjoining sides as clamping surfaces		

Fig. 61: Gripping alternatives which are dependent on clamping positions

conveying facilities no machining power is envisaged, so they can be arranged more simply than apparatus in manufacturing facilities. Because of the dependence of component pick-ups on component geometry, particular account must be taken in the standardisation of component pick-up where there is high component mix. Fig. 62 illustrates, using the example of a conveying pallet, alternatives for division in a basic grid which also serves for the collection of pick-up elements. The pick-up elements can be either interchangeable or capable, through their geometry, of allowing pick-up of differing components.

Planning movement mechanisms. Orientation and positioning for pick-up and placing are crucial in determining spatial arrangement of movements to transfer components between production facilities. In addition there are obstacles to movement which are given in the space situation of the production plant. The planning of necessary movement axes therefore starts from the orientation and position of components in the storage, conveying and manufacturing facilities (Fig. 63).

The demand for the simplest possible movement processes requires that starting from the actual planned situation of the production facilities, the spatial factors are arranged so that the smallest possible

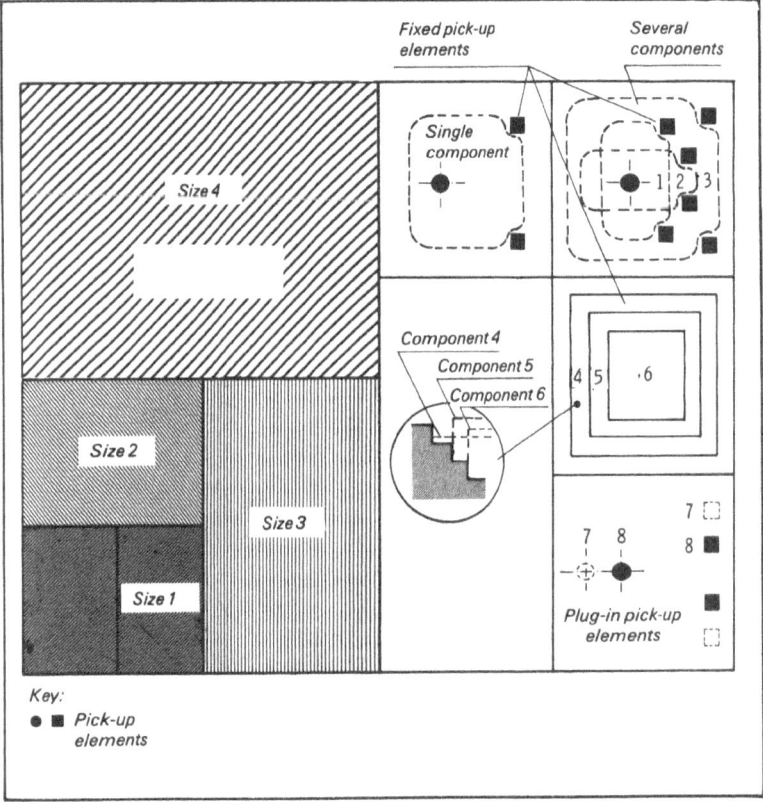

Fig. 62: Configuration of component pick-ups in storage and conveying facilities

number of movement axes is required. By building on this, facilities can be integrated to allow the handling sub-operation 'movement' in production facilities.

Thus, first of all, the movement axes existing at any given time are used (e.g. tool change system, advance, conveying movement, and so on). Since it often follows that the existing movement axes are not sufficient to carry out all necessary movements, it must be investigated how far an expansion of existing movement possibilities with regard to range or turning angle and an increase in constraining tolerances would improve movement processes. Only then can additional movement mechanisms for storage conveying and manufacturing facilities be considered.

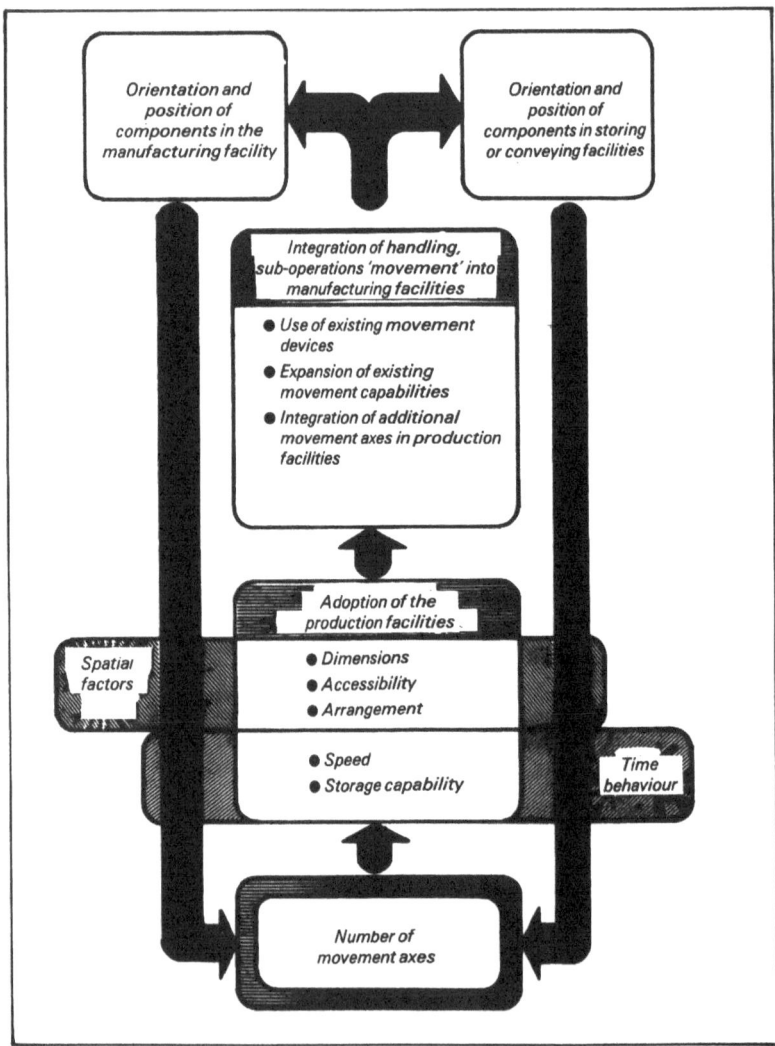

Fig. 63: Determination of movement axes for the performance of handling operations in the supply of manufacturing facilities

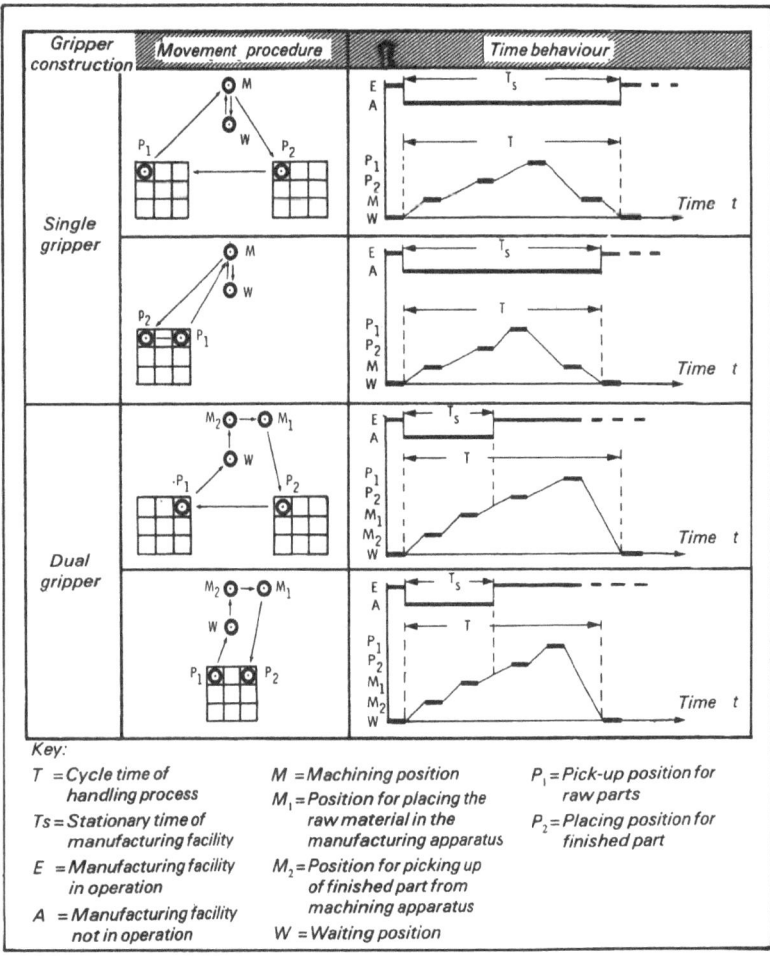

Fig. 64: Effect of paths on the time behaviour of handling facilities

With the direct supply of several manufacturing facilities through a conveying or storage system, a division of movement axes is usefully undertaken. To do this, those handling axes must be placed in storage or conveying facilities which are repeatedly necessary in all manufacturing processes. In particular manufacturing facilities only the additional necessary movements are undertaken by operating units of the machining facilities.

Next to the spatial arrangement of movements, time behaviour of the

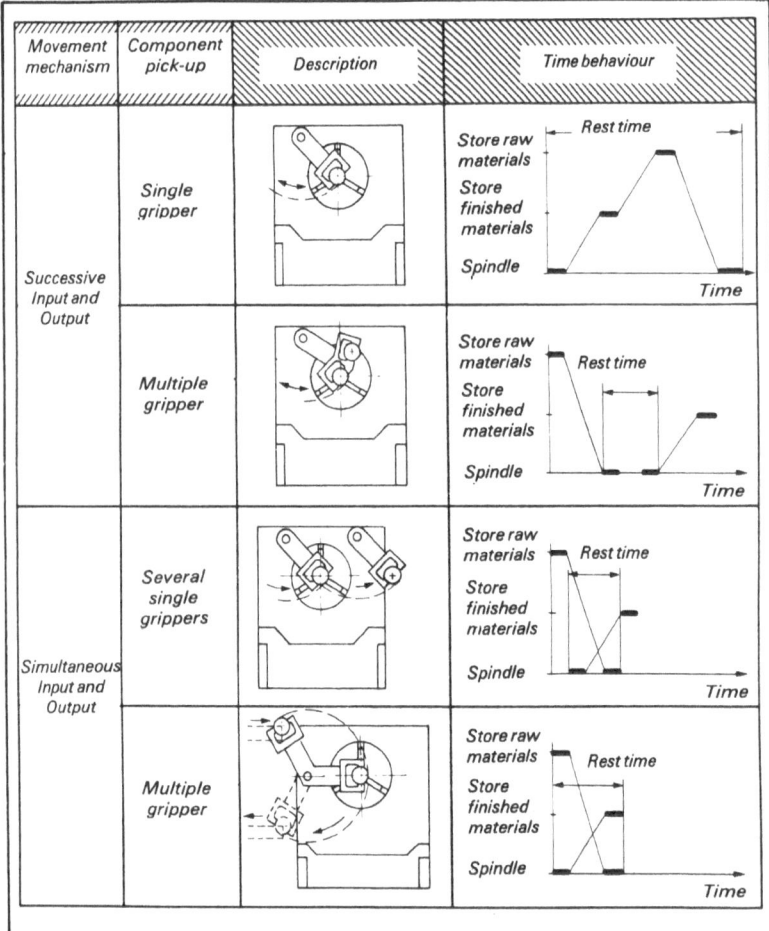

Fig. 65: Effect of carrying out the handling sub-operation 'movement' on the time behaviour of handling facilities

handling facilities is a fundamental part of the planning of movement mechanisms. The objective here must be to minimise the stationary component handling times in production facilities.

Crucial factors in the time behaviour of handling movements are achievable speeds, acceleration and waiting time (e.g. switching times at the beginning of a movement). Through the combination of several movement axes in a production facility, the time behaviour will be

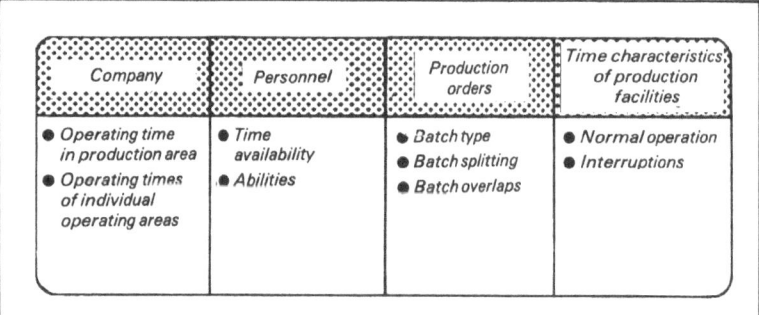

Company	Personnel	Production orders	Time characteristics of production facilities
● Operating time in production area ● Operating times of individual operating areas	● Time availability ● Abilities	● Batch type ● Batch splitting ● Batch overlaps	● Normal operation ● Interruptions

Fig. 66: Criteria for determining store capacity

conclusively determined by whether the movement axes are sequential or simultaneous, since simultaneous movements achieve considerable time savings.

The time savings arrived at through increases in speed and acceleration depend on the paths to be laid down. From Fig. 64 this effect results from distances between positions for pick-up and placing.

Further possible influences on movement in the operational procedures of handling can be altered by collecting together operations, operating units and operational objectives as well as the division of operations among several operating units (see Fig. 18).

Starting with the simplest form of movement, e.g. the supply of manufacturing facilities by a single movement mechanism, the movement operations during insertion and output can be summarised (Fig. 65). The existence of a multiple component pick-up which can store a component for insertion during the output of the machined component is assumed here.

The division of operations among several operating units can be achieved through separate movement mechanisms. In this way the movement mechanisms are synchronised to the extent that simultaneous insertion and output of the component can take place. It is essential here that sufficient movement space is created for both separated movement mechanisms.

Further, the times required for handling are affected by the simultaneous handling of raw and machined materials using a single movement device. Thus the movements for insertion and output are carried out by a single movement mechanism which simultaneously achieves the position for pick-up and placing of components. This movement mechanism must be fitted with two component pick-ups which, because of the limitation at any given time on the insertion and output of components, can conform to the existing raw and finished materials.

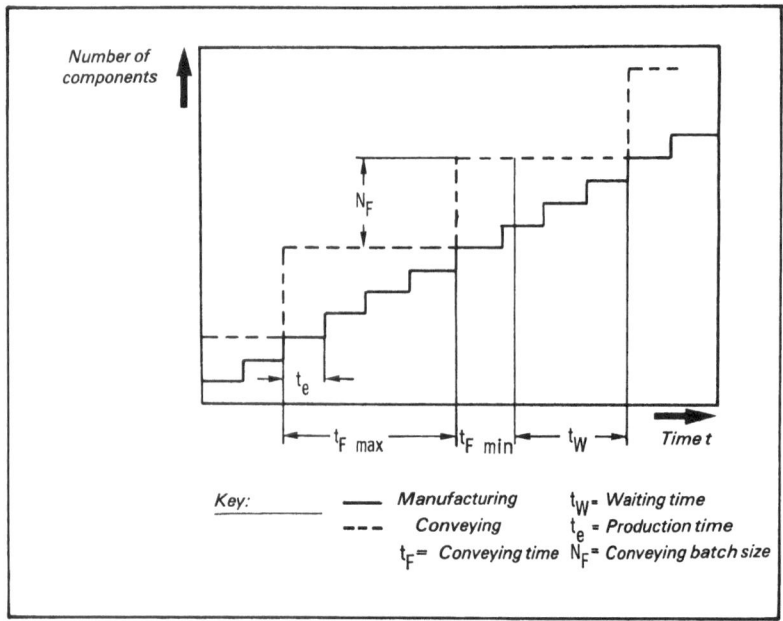

Fig. 67: Determination of conveyed batch size for direct supply of
manufacturing facilities

The movement mechanisms represented in Fig. 65 with rotary
movements can also be achieved with transverse movement axes or with
the aid of combinations of rotary and/or transverse movement axes.
This kind of movement equipment is not linked to the manufacturing
facilities discussed in the previous chapters.

Planning store facilities. Planning store facilities in terms of integrating
handling operations in production plants involves store facilities which
are connected with manufacturing or conveying systems, and which are
quite distinct from the storage function of warehousing.

The construction of store facilities will determine both the type of
store and its capacity.

The type of store facility is established mainly by component-specific
characteristics (e.g. weight, dimensions), the space situation in the
existing production plant and accessibility for movement devices and
component pick-ups. This is no different from the otherwise usual
layout of warehousing in terms of technical investment planning. In this

connection, the planning of dynamic stores in which components are moved is an integral part of planning movement axes as discussed in the previous chapter.

The spatial extent of stores must correspond to the relevant position proportions in conveying and manufacturing facilities. It is taken that:

$$N_{\text{max zul}} = F_L H_L \varkappa_L / V_{WS}$$

where $N_{\text{max zul}}$ is the maximum number of stored components, F_L is the store floor area, H_L the store height, and V_{WS} is the storage volume of components. The storage volume of components corresponds to the geometric dimensions of components plus standard margins, which are necessary as gaps between stored objects to allow satisfactory gripping of components.

The space utilisation level $\varkappa_L$ (space utilisation level = [number of stored components × store volume]/[store floor area × store height]) is dependent on: the arrangement of components in the store (interlacing), the number of store levels (stacking), and the space required for bearing supports, drives (in dynamic stores), component pick-ups etc.

Determination of store capacity involves establishing the number of components (component carriers) to be picked up in a store. Under the heading of 'store capacity' is included the time interval necessary for machining or for distribution of stored components. Store capacity is determined by plant-related, personnel-related and order-related factors, as well as by time characteristics of production facilities (Fig. 66).

The store capacity must principally reflect the division of shifts in which conveying and manufacturing facilities can be ready to work independently from other production facilities in the same area. The store must be able to pick up half, whole or several shift supplies. In order not to be too closely connected with the company's shift divisions along with conveying and manufacturing facilities, it is advisable to over-size the store slightly; in this way no stoppages will be caused in the production facilities through the change-over of personnel between shifts.

Similarly, the operating times of individual operating areas which supply or remove materials can be referred to in the determination of store capacity. If, for example, the raw material suppliers work a single shift, while production operations take place in several shifts, then a storage capacity suitable to this time difference should be provided in the production area. If large material quantities are involved, then the facility of a free-standing store may be necessary. This must be determined within the total store plan.

Personnel is another factor affecting store capacity. This is particularly so in the case in production areas with operating times having reduced numbers of personnel, in which store facilities can compensate for the low availability of operating personnel. In normally manned shifts, the store serves to separate operating personnel from the manufacturing facilities, when supply and loading processes are summarised, so that personnel can undertake other functions, e.g. control of manufacturing output.

A third category of factors affecting the construction of store facilities is production orders. Here the differing types of finishing for production orders must be identified. While in batch-type production, as in batch-splitting, the store is laid out only for the number of components to be machined at any given time, several component types must be envisaged in the determination of store capacity in production facilities with overlapping batches. In particular, production with overlapping orders demands that the supply of store facilities can be carried out independently of component removal, to avoid frequent stoppages in conveying and manufacturing facilities.

Store capacity can be considered from a company, personnel and order-related viewpoint in a clearly static calculation. Roughly speaking this is also possible for layout according to the time characteristics of the production facilities. However, in particular with cost-intensive stores and component carriers (e.g. machine pallets) and with capital-intensive orders, it is necessary to use dynamic methods to determine store capacity, because of crucial time relationships among production facilities.

Whether through static or dynamic considerations, the time differences which occur during normal production between conveying and manufacturing facilities must be involved in determining store capacity. Operational interruptions must also be taken into account.

With direct supply of manufacturing facilities by conveying facilities, the store capacity of the conveying device will involve the conveyed batch size (Fig. 67). This can be derived with the help of the formula:

$$N_F = t_{Fmax} N_B / t_e$$

where N_F is the conveyed batch size, N_B is the number of components to be machined simultaneously, t_{Fmax} is the maximum time duration until the delivery of a further batch, and t_e is the piece time.

Thus the factors N_F and N_B are non-dimensional. For the values t_{Fmax} and t_e, the dimensions are optional, however it is advisable to adopt the usual time designations according to REFA (German Committee for Consideration of Work Times) (minutes).

For indirect supply (machine-linked store), continuous and discontinuous conveying systems are differentiated. In continuously working conveying systems (e.g. suspension chain conveyors), the conveying frequency is constant. As in Fig. 68, the conveyed batch size can be derived from the following formula using the above-mentioned factors:

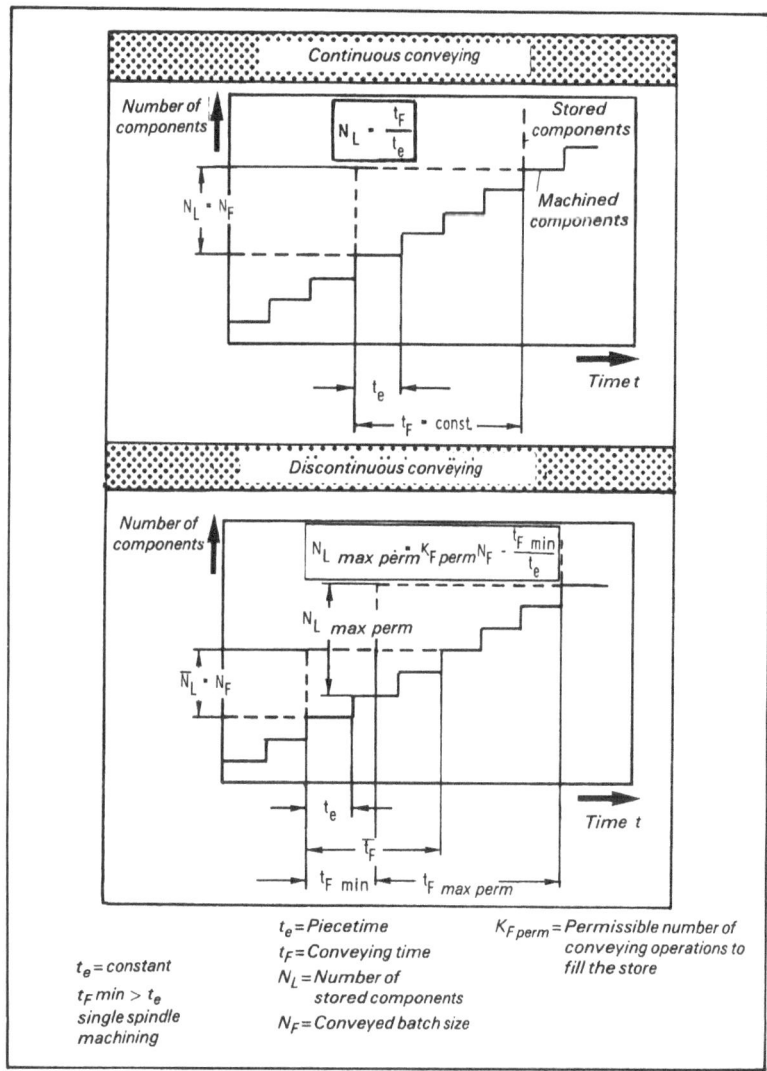

Fig. 68: Relationship between machine-linked store capacity, conveying and piece times

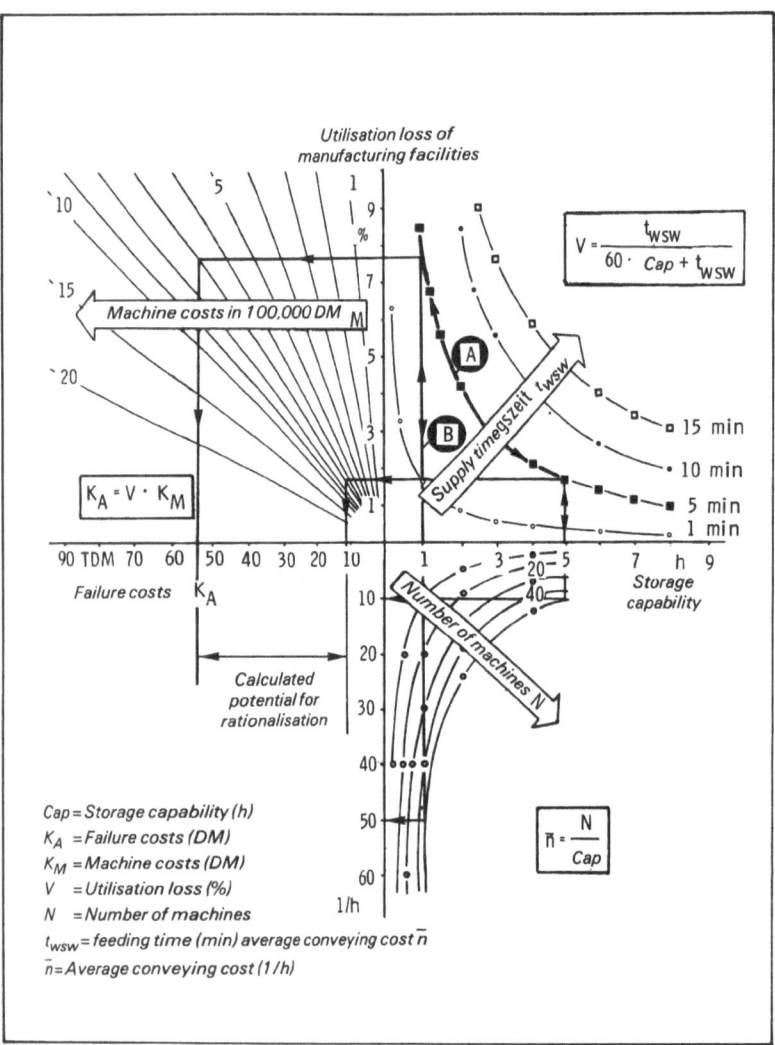

Fig. 69: Effects of the store capacity of machine-linked stores on the loading of manufacturing and conveying facilities

$$N_F = t_F . N_B / t_e$$

Here the conveyed batch size is also the store capacity of the manufacturing facility. From this, it must follow that no residual stocks of components stay in the store of the manufacturing facility. If this is the case, these must be accounted for in the store capacity. This may be necessary, e.g. in the case of stoppages.

In discontinuous conveying, the maximum duration of conveying time t_{Fmax} for the consideration of machine-linked store capacity is to be calculated:

$$N_L = t_{Fmax}/t_e + \text{residual stock}$$

where N_L is the number of stored components.

Because of the discontinuous conveying process, a residual stock of components in the store must be accounted for. The size of this stock depends on the quality of conveying process control. The worst situation is when a sequence of conveying processes with minimum conveying time t_{Fmin} has started.

For the median conveying time t_F it follows that:

$$t_F = (t_{Fmin} + t_{Fmax})/2$$

if an equal distribution of conveying times is to be considered.

For normal operations with balanced component stocks, i.e. neither counting starting nor running down of a production process, the median volume for stores at manufacturing facilities can be derived:

$$t_F = 2N_F t_e/2$$

From this it follows that:

$$N_L = t_F/t_e = N_F$$

Thus the conveyed batch size N_F is determined at the same time, since with balanced component stocks, the production system can be treated as quasi-static with continuous materials flow.

The effect of machine-linked stores on indirect supply for manufacturing facilities is illustrated in Fig. 69. First of all the dependence of the conveying frequency with an increasing number of manufacturing facilities and the utilisation loss of manufacturing facilities is illustrated, which can occur through possible stoppages of supply from the store. As shown, both factors reduce with increasing store capacity, so that this increase not only reduces the utilisation loss of a manufacturing facility but also the conveying frequency (Case A).

An increase in store capacity is, however, only possible to a limited extent (see Fig. 68). The costs associated with utilisation loss do, however, suggest a potential rationalisation, which can also be used, e.g. to accelerate transfer processes in store supply (Case B). Since store supply is principally a question of handling processes, these means are available, especially to produce component pick-ups and movement mechanisms.

The methods described in the early part of this chapter for the development of production concepts with integrated handling operations can be applied in principle to any of the production systems in the investigated field. Thus for different production conditions, independent production concepts may be developed, which are, however, independent from each other and therefore do not have compatible elements.

The development of this kind of special solution is possible only in a few cases in terms of producers and users of production facilities with integrated handling operations, since their achievement will post a considerable outlay of time and finance.

In the following sections, therefore, a concept for a component store with integrated handling operations is proposed, with the aid of which widely differing production systems can be established without a company-specific single solution emerging.

Construction of a component store

The integration of handling operations of storage, conveying and manufacturing facilities in a concrete production system assumes that the production processes involved can be automated. However, in many companies operating single unit and volume production, this is only possible in a limited way in respect of production stores and conveying facilities. The cause of this lies in the customary small conveying volumes and low conveying frequency found in this field, which does not guarantee an economic loading of store or conveying facilities, although automation of these processes is technically feasible.

A further reason for this lack of automation in storage and conveying facilities in single unit and serial production is the failure of specially adapted methods of control for materials flow, necessitated by constant intervention of store and conveying personnel. Considerations of integrating handling operations in production plants involving single unit or serial production must therefore follow from the fact that, as in Fig. 57, it is principally production cells of single or several manufacturing facilities which can be automated.

Machine-linked component stores are a significant feature of production cells. Because of the possible variety of manufacturing facilities which can, if necessary, come into operation in production cells, the integration of handling operations must be particularly concerned with the configuration of the component store. To ensure the widest possible application of such a component store, a building set can be developed out of which the varied forms of component stores for differing production cells can be derived.

Configuration requirements. The component store in production cells is situated at the transfer point between manufacturing facility and conveying system. Thus it needs to fulfill the requirements of both (Fig. 70).

That means in the first instance, that the store is provided with the usual commercial size component carriers, e.g. Europallets (800mm × 1200mm) or industrial pallets (1000mm × 1200mm). These component carriers are of standard dimensions usual in manufacturing industry, so that an automation of handling operations in manufacturing facilities does not necessarily involve a change in store and conveying facilities. The components are ordered in the component carriers for presentation to the component store. To this end they are equipped with positioning elements to prevent unwanted changes in component orientation and position. Thus a component order which is produced, e.g. in other manufacturing facilities before, can be maintained until the components are needed in the production cell, so that the use of operating personnel to line up the component carriers can be dispensed with.

Furthermore, the store is to be set up on the conveying device side for manual supply so that the manufacturing process is not constrained by the supply process. In the case of automated conveying processes a cross-over point to the conveying system must also be created.

To reduce the number of conveying processes, more units of stored component carriers can be changed at each stage of component change.

On the manufacturing facility side, the store must be able to insert components directly into the component pick-up and remove them from there. This necessitates an appropriate accessibility of the component pick-up which is not always given in existing manufacturing facilities. Therefore, the store must also be designed to supply machines with the aid of free-standing flexible handling devices.

The size of the store must depend on the duration of time in which the manufacturing facilities can be operated without manual intervention, as well as on the dimensions of stored components. For incorporation in the largest possible number of potential production systems, the component store should be adaptable for as many combinations of manufacturing facilities and conveying systems as possible. In addition, the component store must be capable of expansion in order to fit differing sizes of production systems. This expansion capability should first of all allow the possibility of expanding automated production systems in stages, starting with units capable of only small operations which by addition of manufacturing facilities, conveying devices and component stores can be broadened into larger systems. Thus a space-saving configuration of the store facilities is usually desirable.

Establishing the elements. Basically a component store can be designed in a dynamic revolving form or in a static form. To store the above-mentioned component carriers, however, revolving stores represent technically costly solutions, since in every store process the total complement of components must be moved. In addition, free accessibility to individual component carriers in any kind of sequence is not possible so that, for example, during component carrier change, the machining process must be interrupted when the change process cannot be finished during the machining of a component actually in the machine tool. These kinds of change processes could be accelerated by exchanging the whole component store, although that would mean additional exchange stores.

Therefore the solution will be a static store. The use of shelving is preferable because of the better utilisation of floor space than a single-level store.

The building elements for the store shelving consist of elements for load pick-up, filling frames and upright staging.

Load pick-up can be designed with compartment bases, mountings, rollers or moving compartments. Compartment bases are particularly recommended for light shelving constructions, especially for the increased rigidity they allow in the total structure. This is especially so if they are designed to be used not only for component carrier pick-up but also to brace and support the staging.

Mountings represent the simplest construction possibility for load pick-up, however, they offer hardly any advantage concerning the firmness of the shelving. In addition they cannot, like compartment bases, actively support the removal process.

This is possible with rollers, by means of which the component carrier can be moved in the shelving. This assumes there is the possibility of driving the rollers. All rollers in a shelving system may be driven separately or centrally using couplings. For a large number of store compartments it is, however, an expensive solution to set up an external coupling drive (e.g. friction drive) for input and output of component carriers.

The use of stub rollers minimises the danger of damage to the shelving by fork-lifts. Alternatively a chain conveyor may be used instead of the rollers.

Filling frames can be fixed, stackable or transportable. The fixed version is suitable for stationary stores in proximity to machines. For components which are brought to the machines in large piece numbers (high conveying volume), a stackable store is recommended because several component carriers can be conveyed simultaneously on these so that the conveying frequency can be reduced along with the number of stored component carriers. The use of stackable filling frames assumes

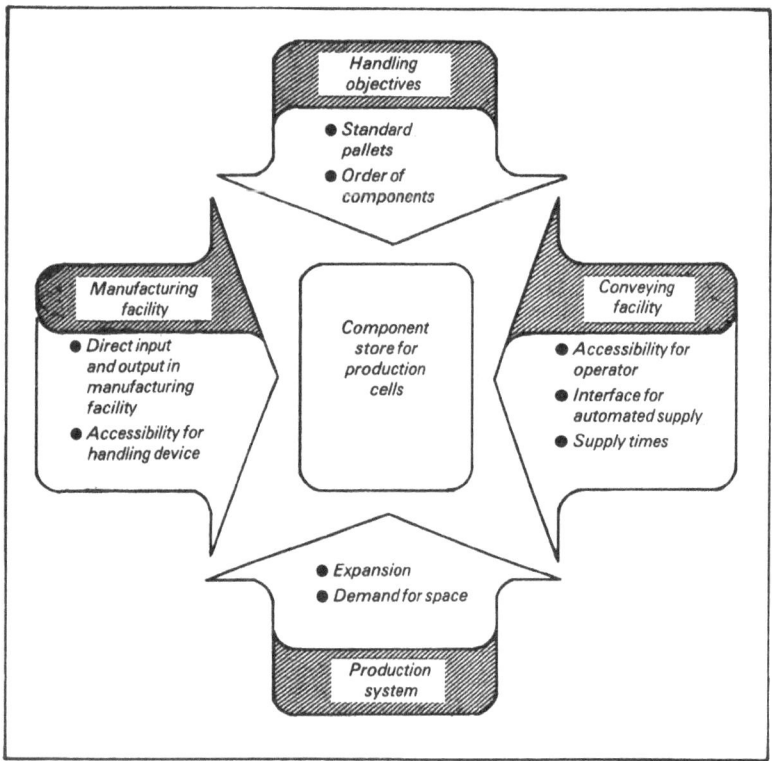

Fig. 70: Requirements for the configuration of a component store with integrated handling operations

pick-up elements in the upper part of the shelving.

The use of transportable and stackable filling frames offers the possibility of fast changeover of the components in a production cell by the exchange of a complete store unit. With transportable stores an independent conveying device is not necessary here (manual shifting).

In choosing the type of filling frame, the smallest possible structure height must be considered so that the store stays as low as possible, otherwise, in certain circumstances, crane runways would be hindered above the store.

Staging is differentiated mostly according to length. The length is determined by:

● The required store capacity (number of component carriers).
● The height of the component carriers and load pick-ups.

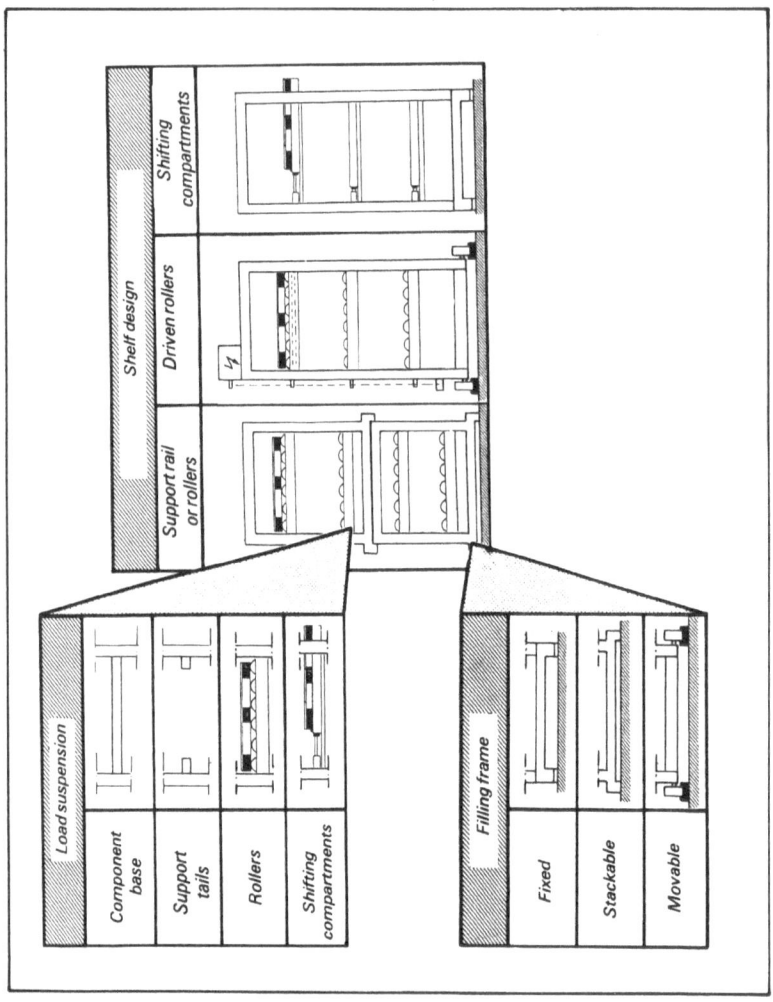

Fig. 71: Building elements of store shelving

- The height of stored components; with direct accessibility within the store, the movement space of the handling device should also be considered.
- The height dictated by crane runways, roof construction, cable runs, etc.

To increase store capability, several shelving sets can be arranged next to one another.

On the shelving on the conveying system side there can be direct manual accessibility, if, for example, single components should be removed for monitoring manufacturing output. This assumes shelf sizing does not exceed the reach of service personnel. Here also a suitable clearance height must be selected for the store compartments. The total height of the shelving and the position of the lowest shelf compartment should be determined according to the ergonomic requirements of service personnel.

On the other hand the shelving may be accessible on the conveying system side to the usual conveying systems e.g. fork-lifts, so that the component carriers can be inserted directly into the shelving compartments. The direct accessibility assumes, however, a clear arrangement of store compartments and component carriers, otherwise some component carriers may get out of order.

On the manufacturing facility side, in the first instance a free-standing handling device can access the components directly. This direct accessibility is, however, problematical in that at the present time there is hardly a device on the market which has the reach and movement capabilities to reach into such a store and grip all potentially possible component positions within a component carrier.

In addition, the low height of the workspace of the usual handling devices would allow only a small number of shelving compartments to be arranged on top of each other. Direct accessibility to components requires reference marks for component carriers in the shelving compartments, so that the positioning and orientation of components is secured within the tolerance range of the component pick-up of the handling device.

To increase the store accessibility, a service device (distribution vehicle) can be used. The building elements for this vehicle consist of elements for the load pick-up, base and lifting apparatus (Fig. 72).

As with store shelving, the load pick-up consists of mountings, driven or non-driven rollers or chain conveying facilities. Pick-up or placing of component carriers, for example, in shelving with mountings as load pick-ups, can be achieved by additional telescopic forks. These telescopic forks also facilitate the horizontal shifting of component carriers so that the components on the carriers can be positioned in this direction.

A change of component orientation is possible by swivelling and turning apparatus in the load pick-up. The base can be constructed as a stand unit with and without turning stands or as a transport unit. With the aid of the transport unit, the distribution vehicle can not only access several shelving units but also supply several manufacturing facilities.

The lifting equipment can be driven electro-mechanically (chain

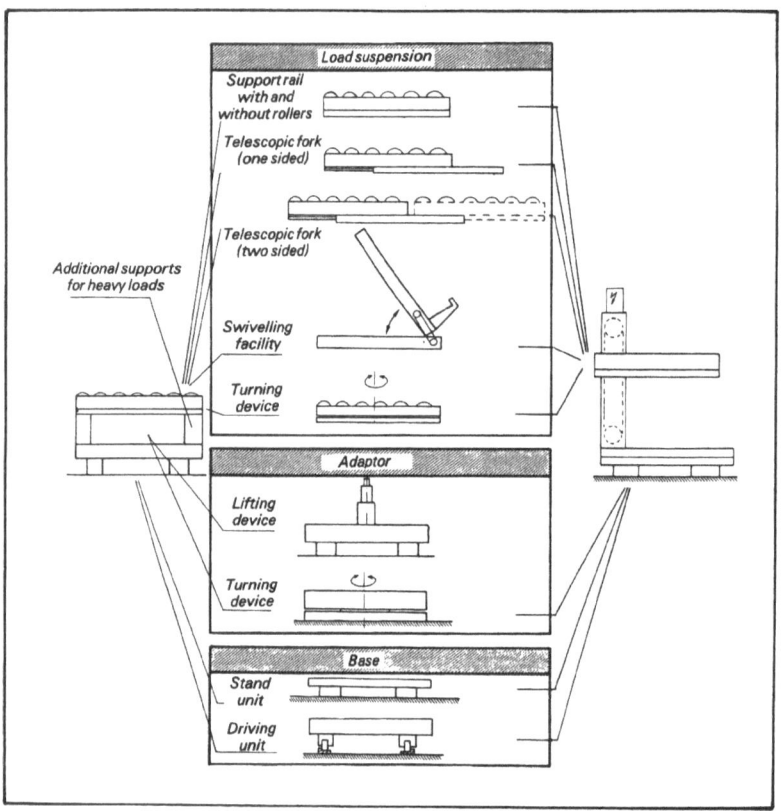

Fig. 72: Building element system for waiting stations and distribution vehicles

drive, scissor lifting table with shaft) or hydraulically. The lifting height will depend on the store height.

For precise positioning of the distribution vehicle in horizontal and vertical directions there is the possibility of using NC-axes with their own odometrical system. In this way constant supervision of the position of the distribution vehicle is possible. This is, however, only necessary in a few cases, since the distribution vehicle needs to move only to constantly repeated discrete positions. It is advisable to introduce markers for these positions, which can be picked up by a simple sensor system. The precision of positioning is then limited to specific areas.

This can also be done for changing positions, as, for example, in component transfer to cantilever milling machines with movable mounting surfaces.

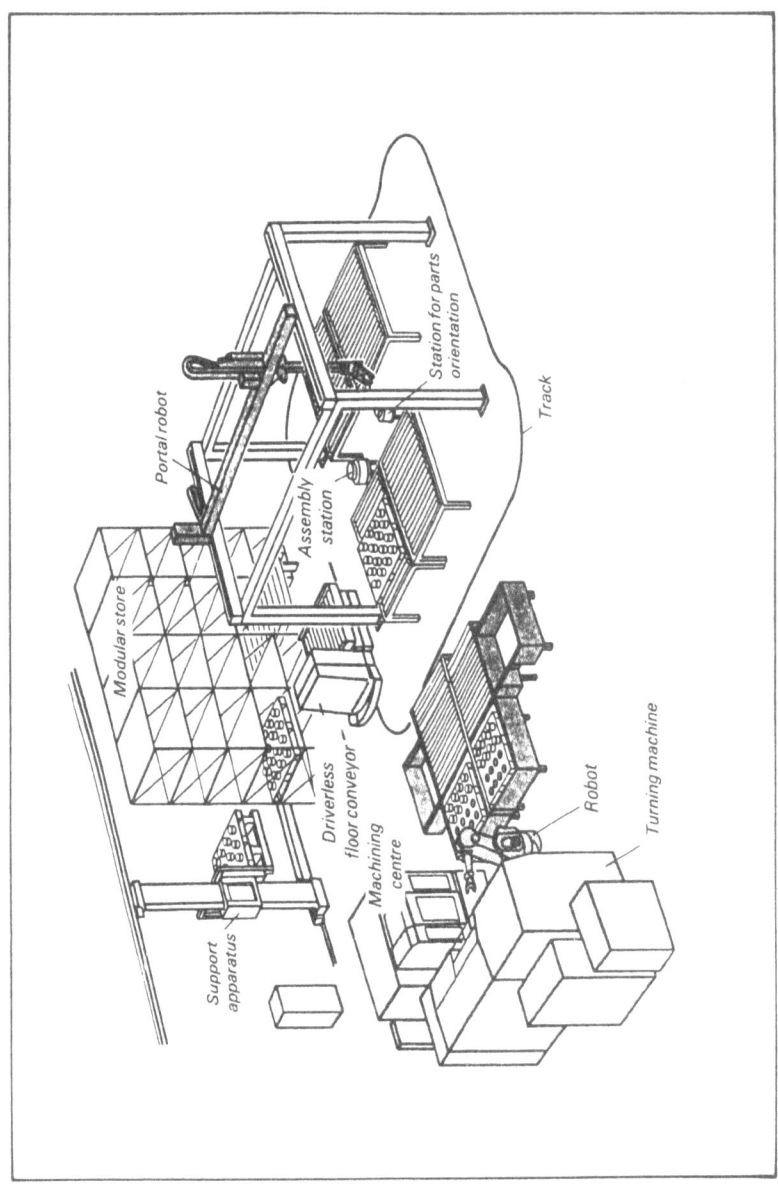

Fig. 73: Component store for a flexible production system

Here the markers must move in conjunction with the mounting surfaces and thus stay in the measuring area of the sensors.

In order to be able to present components with short machining times to several manufacturing facilities, waiting stations are necessary close to the manufacturing facilities. The building elements for waiting stations consist of elements for load pick-up, an intermediate part and a base. The elements for load pick-up and base are identical to those for distribution vehicles.

The intermediate part is fixed in most cases. By equipping it with lifting and turning apparatus, it can, however, bring components into position for the final transfer to the manufacturing facility.

In addition, the waiting station, in conjunction with the distribution vehicle, can be used to carry out the transfer between conveying system and component store. Here the waiting station in the area of the conveying system is accessible from one side for manual supply by operating personnel or for supply through a manually controlled or automatic conveying device, and from the other side is served by the distribution vehicle. The distribution vehicle then takes over the conveying process between the waiting station and the shelving or the manufacturing facility.

An example of the application of building elements for component stores is illustrated in Fig. 73.

This illustration is a concept for a flexible production system to manufacture housings and axles. Two manufacturing facilities (turning and milling) are supplied with the aid of a flexible handling device. The production system is supported by an inductively-driven floor conveying vehicle with palletted components. The components are transferred from the conveying vehicle to a waiting station and from there are loaded into a shelving store.

From there they can be collected again and prepared for the handling device.

In addition to the manufacturing facilities, an assembly cell belongs to the flexible production system, in which the waiting station elements can be similarly used.

Chapter Seven
Conclusion

THIS study has been concerned with investigating the possibilities for integrating handling operations in storage, conveying and manufacturing facilities. Building on a functional consideration of production processes, technical alternatives for performing the handling sub-operations 'orientation, position, quantity, sequence and time order changes' were systematically developed. On the basis of these alternatives, methods of automating production systems with the aid of production facilities with integrated handling operations have been illustrated.

Thus, in the first instance, the requirements of automatic handling on the part of the objects to be handled (components, component carriers) and on the part of the various storage, conveying and manufacturing facilities were described. Special weight was given to the investigation of the space situation in the production facilities considered and their time characteristics.

To describe the procedures involved in the production processes, the operations of storage, conveying and manufacturing systems were analysed and a functional model was developed on the basis of this analysis. This functional analysis necessitated the clarification of the terms 'system', 'system element' and 'operation'. On the basis of these theoretical system relationships, production systems, i.e. the storage, conveying and manufacturing systems of production, were studied. The analysis of elements, operations and sub-operations of handling systems formed the foundation for developing alternative methods of integrating handling operations in production facilities.

It was shown that the total operation of 'handling' could be classified into the operations of 'position change, orientation change, sequence change and quantity change'. Furthermore, it can be the function of the handling system to influence the timed appearance of a component, in that handling procedures are performed discretely or are synchronised with other production processes.

The individual handling operations can be described in terms of a repeating sequence of the handling sub-operations 'pick-up, move, sort and place'.

A comparison of handling operations with the procedures occurring in production facilities shows that in storage, conveying and manufacturing facilities a series of operations are performed, which can be used for the automatic handling of components.

Building on the functional analysis of production processes and the analysis of requirements for automated handling, alternative solutions for storage facilities with integrated handling operations could then be illustrated. To this end the alternative operational procedures possible in storage facilities were systematically presented. The operating units involved with these procedures in the performance of handling sub-operations 'pick-up, move, store and place' were then established.

The measures for integrating handling operations in conveying facilities could be derived as for storage facilities. In addition to technical equipment, the configuration of the conveying track can be used in conveying systems to perform handling processes.

In principle the same procedures could be used in the integration of handling operations in manufacturing facilities. Because of the complexity of manufacturing facilities, there is a multiplicity of technically possible solutions here. These can, however, be summarised as measures:

• to integrate additional component pick-ups,
• to use and extend the movement possibilities of existing operating units, and
• to integrate additional storage positions.

according to the handling sub-operations to be performed.

With the aid of the technical alternatives for integrating handling operations in storage, conveying and manufacturing facilities, production systems could be developed in which these kinds of production facility were included.

Since for technical, economical, personnel and organisational reasons the automation of all production facilities in a production system by the integration of handling operations is not sensible, it is important first of all to define production areas which are appropriate for automation by the integration of handling operations in production facilities.

By building on the definition of production areas, the handling facilities for pick-up/placing, moving and storage of components can be proposed.

The nature as well as the behavioural characteristics of components affects the configuration of component pick-up. Furthermore, the space situation in production facilities and their timing characteristics affect the alternative methods of performing the handling sub-operations 'pick-up' or 'placing'. Thus the number of components to be picked up by a production facility must be initially determined. On the strength of this the component pick-up can be set up according to physical principles and dimension, gripping power, precision, etc. Finally, the pick-up elements can be set up as cross-over points between the components and the component pick-ups.

In planning the necessary movement axes for materials handling, the spatial factors in production facilities should so correspond with each other that the least possible movement axes are necessitated.

Following from this, devices for moving components can be integrated into production facilities. On the one hand, existing movement possibilities can be used or expanded and on the other hand additional movement axes can be integrated into production facilities.

Next to the spatial arrangement of movements, the movement mechanisms also determine for the most part the times required for component handling. To influence the times required for handling processes, not only can the paths be shortened, but the movement speeds and accelerations can be increased. Also a change in handling times can be brought about by dividing the movement procedures among several operating units.

The task of planning store facilities is not only to determine the type of store but to establish store capacity. While the store type is determined principally by the geometrical and technological characteristics of the components and production facilities, operational, personnel and order-related factors can affect store capacity. These factors are used to determine store capacity in static terms. For a dynamic determination of store capacity, the timing characteristics of production facilities must be taken into account.

This type of dynamic calculation or simulation is, however, relatively costly. In terms of the study, the basic relationships for a rough estimate of store capacity on the basis of static values were illustrated.

Because of the large applications range of production cells, this study has proposed the outline of a component store with integrated handling operations which can be set up in production cells. To this end a building block system was developed, with the help of which the component store can be adapted for existing production facilities and conditions.

References

[1] Dronsek, Dunkler and Etzenbach, 1978. Operational possibilities and limits of automated production concepts. *Ind. Auz.*, 100 (73): 38-44.

[2] von Poblotzki, J., 1980. Improvements in the use of time in manufacturing centres. In, *Milling '80.* Fritz Werner, Hamburg.

[3] Junghanns, W., 1971. Planning new production systems for single unit and serial production. TH dissertation, Aachen College of Technology.

[4] Babel, W., Bauklof, K.-F., Betsoh, H., et al., 1981. Machine tool concepts - State and technology. In, *17th Machine Tool Colloqium*, pp. 132-151. Laboratory for Machine Tools and Production Engineering, Aachen.

[5] Adam, P., Armbruster, N., Averkamp, T., et al., 1981. Measurements in processes - Increase in automation/quality stability. In, *17th Machine Tool Colloquim*, pp. 185-199. Laboratory for Machine Tools and Production Engineering, Aachen.

[6] Westkämper, E., 1977. Automation in single unit and serial production. TH dissertation, Aachen College of Technology.

[7] Franzius, H., 1978. Consideration of some major views on store planning. *J. Industrial Production*, 68 (11): 699-705.

[8] Grodula, E., 1973. *Basics of Materials Economy.* Gabler-Verlag, Wiesbaden.

[9] Martin, H., 1979. *Materials Flow and Store Planning*, Springer-Verlag, Berlin.

[10] Jünemann, R., 1971. *System Planning for Piece Goods Store.* Krausskopf-Verlag, Mainz.

[11] Hausmann, G., 1972. *Automated Stores.* Krausskopf-Verlag, Mainz.

[12] Schweizer, M., 1983. 1982: A look back on a robot year. *Production*, 1: 3-4.

[13] Wiendahl, H.-P., 1982. Technical investment planning. TH dissertation, Aachen College of Technology.

[14] Wiewelhove, W., 1977. Systematic data preparation for investment planning. *VDI Journal*, 119 (24): 1198-1206.

[15] Aggteleky, B., 1970. *Factory Planning.* Hauser-Verlag, Munich.

[16] Baur, K., 1972. *Working Capital Organisation in Factory Planning.* Krausskopf-Verlag, Mainz.

[17] Olbrich, W. and Wiendahl, H.-P., 1972. Methods of investment planning for single unit and small series production. *J. Industrial Production*, 62 (11): 246-248.

[18] Setzer, H. and Möllers, K.-H., 1979. *Guide to the Choice of Conveying Devices.* Berlin.

[19] *Component Handling in Transfer Lines without Component Carriers.* VDI-Richtlinie 3238. VDI-Verlag, Düsseldorf, 1968.

[20] *GF Automation.* Customer information from the George Fischer Company, Schaffhausen.

[21] *Automatic Component Handling.* Gildemeister Report 5. Customer information from the Gildemeister Company, Bielefeld.

[22] Töushoff, T., 1982. Feeding apparatus for cylinder operated machines. *J. Industrial Production*, 72: 197-202.

[23] Warnecke, H.-J. and Weiss, K., 1978. *Catalogue of Feeding Apparatus.* Krausskopf-Verlag, Mainz.

[24] Boothroyd, G., Poli, C. R. and Murch, L. E., 1980. *Handbook of Feeding and Orienting Techniques for Small Parts.* University of Massachusetts, Amherst.

[25] *Vibration Technology.* Parts conveyors and building blocks for assembly and handling technology. Customer information from AEG, Berlin, 1979.

[26] Schimke, E. F., 1976. Planning handling systems. TH dissertation, Aachen College of Technology.

[27] Warnecke, H.-J. and Schraft, R. D., 1973. *Industrial Robots.* Production Technology Today (Series), Vol. 4. Krausskopf-Verlag, Mainz.

[28] Stensloff, H. (Ed.), 1980. *Methods for Very Advanced Handling Systems.* Springer-Verlag, Berlin.

[29] van Brussel, H., Engel, G., Eversheim, W., et al., 1981. Flexible automatic handling in production and assembly. In, *17th Machine Tool Colloqium*. Laboratory for Machine Tools and Production Engineering, Aachen.

[30] Engel, G., 1980. Plans and layout for modular handling systems. TH dissertation, Aachen College of Technology.

[31] Spur, G., Severin, F. and Maier, G., 1981. Development of flexible feeding devices for NC-lathes with integrated measuring system. In, *Proc. 6th Int. Conf. on Production Research*.

[32] *FX-RBT Integrated Component Manipulator for the Humane Workplace*. Customer information from Ikegai Co., Japan.

[33] Armbruster, N., 1981. Integration of manufacturing and handling operations. In, *Fabrik '81*. VDI-Verlag, Düsseldorf.

[34] *Automation*. Customer information from Boehringer Co., Cöppingen.

[35] *Check Costs by Higher Productivity*. Customer information from Heyligenstaedt Co., Giepen.

[36] Dietz, P., 1982. Turning by dynamic machining: Loading spindle systems increase productivity. *Maschinenmarkt*, 88 (68): 1377-1380.

[37] Schütz, W. and Steinhilper, R., 1982. Reasonably priced pallet stores for the manufacturing centre. *Industrial Production*, 72 (3): 151-155.

[38] Shimizu, T., 1980. Japan produces faster. *VDI-N*, 48: 6.

[39] Rettler, N., 1982. The application of automated shelving conveyors to link machine tools. *ZWF*, 77 (7): 309-319.

[40] Böckmann, H., 1980. *Revolving and Transverse Shelf Installations Operational Possibilities in Small and Medium Sized Companies*. Beuth-Verlag, Berlin.

[41] Flexible handling and store system. *Technica*, 31 (1): 36-38, 1982.

[42] Gunsser, P. and Harter, W., 1982. Linking mechanical production areas and manufacturing centres with driverless transport systems. *ZWF*, 77 (7): 301-304.

[43] Rössner, W., 1982. Electric suspended rails with automated handling facilities. *ZWF*, 77 (3): 105-108.

[44] New FMW uses robotrailer. *Metalworking, Engineering and Marketing*, September 1981: 114-119.

[45] Handke, G. and Beumer, P., 1979. Radio-controlled storage and transport system supplies cutting/machining production process. *Conveying and Lifting*, 29 (7): 611-616.

[46] Pörsche, M., 1980. Types of store compared for economic value. *Management Journal*, 49 (3): 131-138.

[47] *Terms and Explanations in Conveying.* VDI-Richtlinie 2411. VDI-Verlag, Düsseldorf.

[48] *Continuous Conveyors for Piece Goods. Overview with Judgement Criteria.* VDI-Richtlinie 2342. VDI-Verlag, Düsseldorf.

[49] *Continuous Conveyors.* DIN-Norm 15201, Part 1. Beuth-Verlag, Berlin.

[50] Rössner, W., 1981. Automating materials flow with circular chain conveyors and handling facilities. *ZWF*, 76 (8): 394-398.

[51] Füglein, E., 1981. *Mobile Pallet Stores.* Forschungsberlicht HA 81-022 of BMFT research report. Kernforschungszentrum Karlsruhe (KfK), Karlsruhe.

[52] Herrmann, G., 1976. Analysis of handling processes with reference to their requirements for programmable handling devices in part production. TU dissertation, Stuttgart.

[53] Auer, B. H., 1977. Contribution to increase flexibility of handling facilities in the scope of single unit and small run production. TU dissertation, Berlin.

[54] Bürgel, W., Eversheim, W., Burkhardt, G., et al. Integration of Handling Operations in Storing, Conveying and Manufacturing Systems. A contribution towards automation of production facilities with special reference to the manufacture of prismatic components. BMFT-KfK-PFT-Forschungsbericht, Kernforschungszentrum Karlsruhe (KfK), Karlsruhe.

[55] Rittinghausen, H., 1980. Integrated materials flow automation in single unit and serial production. *Reihe Produktionstechnik Berlin*, Vol. 12. Carl Hanser Verlag, Munich.

[56] Hesse, S., 1981. Component ordering in manipulator-peripheral stores. *Fertigungstechnik und Betrieb*, 31 (10): 611-614.

[57] Spur, G. and Viehweger, B., 1982. Computer-assisted materials preparation in flexible production cells. *ZWF*, 77 (10): 494-498.

[58] Müller, W., 1982. Market analysis of turning machines. Unpublished study – Machine Tool laboratory.

[59] *ABC of Clamping Technology.* Customer information from Forkardt Co., Düsseldorf.

[60] *Methods and Means of Systematic Apparatus Construction.* Study of the VDI committee 'Apparatus'. VDI-Verlag, Düsseldorf, 1983.

[61] Redecker, G. and Janisch, H., 1981. Buffer optimisation in linked conveyor systems. *ZWF*, 76 (12): 579-585.

[62] Petermann, E. and Reinecke, P., 1982. Determination of economic intermediate store structures. *VDI-Z*, 124 (5): 177-182.

[63] Pferdmenges, R., 1980. Organisation in flexible automated production concepts. TH dissertation, Aachen College of Technology.

[64] Hubka, U., 1983. *Theory of Machine Systems.* Springer-Verlag, Berlin.

[65] Müller, W., 1979. Development of new production facilities. *Ind.-Anz.,* 101 (39): 62-63.

[66] Eversheim, W., 1978. Investigation of technical and organisational requirements in the layout and use of automated manufacturing units for discontinuous piece goods production. TH dissertation, Aachen College of Technology.

[67] Thumm, R., 1981. Symbols for operations plans for material and information flow. Report, Institute for Conveying Techniques, University of Karlsruhe.

[68] *Handling Operations: Terms, Definitions, Symbols.* VDI-Richtlinien 2860 (Entwurf), VDI-Verlag, Düsseldorf, 1983.

[69] Wiesner, F., 1980. Flexible component handling and machine linking with portal and pallet systems. *Werkstatt und Betrieb*, 113 (3): 159-165.

[70] Müller, W., 1979. Ordering apparatus for the simplification of handling processes in precision engineering. *Ind.-Anz.,* 101 (82): 25-28.

[71] Eversheim, W., Ehl, R., Hoeschen, R.-D., et al., 1980. Ordering apparatus for the simplification of handling processes in precision engineering. Laboratory for Machine Tools and Production Engineering, Aachen.

[72] Elbracht, D. and Pfeiffer, D., 1982. Inductively controlled industrial robots. *wt.Z. ind. Fertigung*, 72: 689-692.

[73] Peffekoven, K. H., 1982. Configuration of turning machines suitable for materials handling. TH dissertation. Aachen College